Introduction to Modern Magnetohydrodynamics

Ninety-nine percent of ordinary matter in the Universe is in the form of ionized fluids, or plasmas. The study of the magnetic properties of such electrically conducting fluids, magnetohydrodynamics (MHD), has become a central theory in astrophysics, as well as in areas such as engineering and geophysics. This textbook offers a comprehensive introduction to MHD and its recent applications, in nature and in laboratory plasmas; from the machinery of the Sun and galaxies, to the cooling of nuclear reactors and the geodynamo. It exposes advanced undergraduate and graduate students to both classical and modern concepts, making them aware of current research and the ever-widening scope of MHD. Rigorous derivations within the text, supplemented by over 100 illustrations and followed by exercises and worked solutions at the end of each chapter, provide an engaging and practical introduction to the subject and an accessible route into this wide-ranging field.

Sébastien Galtier is a Professor of astrophysics at the Université Paris–Saclay, France. His research focuses on magnetohydrodynamic turbulence, and he has published widely in the field. He was President of the French National Program in Solar Physics (CNRS) during the years 2010–2014, and is an honorary member of the prestigious Institut Universitaire de France.

Introduction to Modern Magnetohydrodynamics

Sébastien Galtier

Université Paris–Saclay

Shaftesbury Road, Cambridge CB2 8EA, United Kingdom

One Liberty Plaza, 20th Floor, New York, NY 10006, USA

477 Williamstown Road, Port Melbourne, VIC 3207, Australia

314–321, 3rd Floor, Plot 3, Splendor Forum, Jasola District Centre, New Delhi – 110025, India

103 Penang Road, #05–06/07, Visioncrest Commercial, Singapore 238467

Cambridge University Press is part of Cambridge University Press & Assessment,
a department of the University of Cambridge.

We share the University's mission to contribute to society through the pursuit of
education, learning and research at the highest international levels of excellence.

www.cambridge.org
Information on this title: www.cambridge.org/9781107158658

© Editions Vuibert – Paris 2013
English translation © Sébastien Galtier 2016

First published 2013 as *Magnétohydrodynamique – Des plasmas de laboratoire à l'astrophysique*
Updated English edition published 2016

A catalogue record for this publication is available from the British Library

Library of Congress Cataloging-in-Publication data
Names: Galtier, Sébastien, author.
Title: Introduction to modern magnetohydrodynamics / Sébastien Galtier,
Université Paris–Saclay
Description: Cambridge, United Kingdom ; New York, NY : Cambridge University
Press, 2016. | Includes bibliographical references and index.
Identifiers: LCCN 2016014636| ISBN 9781107158658 (Hardback) | ISBN 1107158656 (Hardback)
Subjects: LCSH: Magnetohydrodynamics.
Classification: LCC QC718.5.M36 G35 2016 | DDC 538/.6–dc23 LC record available at
https://lccn.loc.gov/2016014636

ISBN   978-1-107-15865-8   Hardback

To my family

Contents

Preface

Physical laws should have
mathematical beauty
P. A. M. Dirac – Nobel Prize in Physics (1933)

In our familiar environment, matter appears in solid, liquid, or gaseous form. This triptych vision of the world was shaken in the twentieth century when astronomers revealed that most of the extraterrestrial matter – namely more than 99% of the ordinary matter in the Universe – is actually in an ionized state called plasma whose physical properties differ fundamentally from those of a neutral gas. The study of this fourth state of matter was developed mainly in the second half of the twentieth century and is now considered a major branch of modern physics. A decisive step was taken in 1942 when the Swedish astrophysicist Hannes Alfvén (1908–1995) proposed the theory of magnetohydrodynamics (MHD) by connecting the Maxwell electrodynamics with the Navier–Stokes hydrodynamics. In this framework, plasmas are described macroscopically as a fluid and the corpuscular aspect of ions and electrons is ignored. Nowadays, MHD has emerged as the central theory to understand the machinery of the Sun, stars, stellar winds, accretion disks around super-massive objects such as black holes with the formation of extragalactic jets, interstellar clouds, and planetary magnetospheres. Also, when H. Alfvén was awarded the Nobel Prize in Physics in 1970, the Committee congratulated him "for fundamental work and discoveries in magnetohydrodynamics with fruitful applications in different parts of plasma physics."

The MHD description is not limited to astrophysical plasmas, but is also widely used in the framework of laboratory experiments or industrial developments for which plasmas and conducting liquid metals are used. In the first case, the emblematic example is certainly controlled nuclear fusion with the International Thermonuclear Experimental Reactor (ITER) in Cadarache. Indeed, the control of a magnetically confined plasma requires an understanding of the large-scale equilibrium and the solution of stability problems whose theoretical framework is basically MHD. Liquid metals are also used, for example, in experiments to investigate the mechanism of magnetic field generation – the dynamo effect – that occurs naturally in the liquid outer core of our planet via turbulent motions of a mixture of liquid metals. Most of the natural MHD flows cited above are far from thermodynamic equilibrium, with highly turbulent dynamics. Furthermore, a finer description including the most important effect, i.e. the decoupling effect between the ions and the electrons – the Hall effect, is nowadays often used to understand observations and experiments. Thus, an introduction to modern MHD must include both turbulence and the Hall effect, which is the case of this book where a systematic comparison with recent research is made with a large number of citations.

This textbook is an introduction to modern MHD. It provides a clear connection between the theory and recent experimental results. It aims at presenting the main physical properties and applications of plasmas or liquid MHD flows starting from the knowledge of an undergraduate student. It is therefore addressed primarily to advanced undergraduate students, postgraduate (Masters) students – regardless of their area of specialization (astrophysics, plasma, fusion, or fluid mechanics), and engineering students wishing to complete their training. Mathematical derivations are rigorous and the results are illustrated with more than 100 figures, some of which originate from the most recent experimental measurements. Exercises with their solutions complete the presentation. Approximately 80% of the content of this textbook corresponds to a one-semester postgraduate MHD course that I give regularly at the Université Paris–Saclay and which was published in French in 2013. The present version is its English translation with some new material.

I am grateful to all my Masters students, PhD students, and colleagues with whom I have discussed MHD, and in particular to Supratik Banerjee, Romain Meyrand, and Caroline Nore.

Paris, 29 August 2015   Sébastien Galtier

Table of Physical Quantities

Numerical values of some (plasma) parameters appearing in the main text. The international system (IS) is used (densities n_e and n_i are in m^{-3}; magnetic field B in tesla (T); temperature T in kelvins; magnitude of the electron charge e in coulombs; electron and ions masses m_e and m_i in kg) and we assume that ions are only protons. In the evaluations of v and η, we consider a completely ionized plasma (Spitzer 1962).

Electron plasma frequency	$\dfrac{\omega_{pe}}{2\pi} = \dfrac{(n_e e^2/m_e \varepsilon_0)^{1/2}}{2\pi} \simeq 8.98\, n_e^{1/2}$ Hz
Ion plasma frequency	$\dfrac{\omega_{pi}}{2\pi} = \dfrac{(n_i e^2/m_i \varepsilon_0)^{1/2}}{2\pi} \simeq 0.21\, n_i^{1/2}$ Hz
Electron inertial length	$d_e = c/\omega_{pe} \simeq 5.3 \times 10^6\, n_e^{-1/2}$ m
Ion inertial length	$d_i = c/\omega_{pi} \simeq 2.3 \times 10^8\, n_i^{-1/2}$ m
Electron gyrofrequency	$\dfrac{\omega_{ce}}{2\pi} = \dfrac{eB/m_e}{2\pi} \simeq 2.8 \times 10^{10} B$ Hz
Ion gyrofrequency	$\dfrac{\omega_{ci}}{2\pi} = \dfrac{eB/m_i}{2\pi} \simeq 1.5 \times 10^7 B$ Hz
Kinematic viscosity	$v \simeq 10^{10}\, T^{5/2}\, n_i^{-1}$ m^2/s
Magnetic diffusivity	$\eta \simeq 10^9\, T^{-3/2}$ m^2/s
Reynolds number	$R_e = Lu/v$
Magnetic Reynolds number	$R_m = Lu/\eta$
Lundquist number	$S = LB/(\eta \sqrt{\mu_0 m_i n_i})$
Magnetic field strength	1 Tesla $= 10^4$ Gauss
Magnetic pressure	$P_m = B^2/(2\mu_0) \simeq 4 \times 10^5 B^2$ Pa
Length scales	1 pc $\simeq 3.2$ light-years $\simeq 3 \times 10^{16}$ m

Part I

Foundations

I

Introduction

1.1 Space and Laboratory Plasma Physics

Physics has experienced several revolutions in the twentieth century that profoundly changed our understanding of nature. Quantum mechanics and (special, general) relativity are the best known and certainly the most important, but the discovery of the fourth state of matter – the state of plasma – as the most natural form of ordinary matter in the Universe, with more than 99% of visible matter being in this form, is unquestionably a revolution in physics. This discovery has led to the emergence of a new branch of physics called *plasma physics*.

Plasma physics describes the coupling between electromagnetic fields and ionized matter (electrons, ions). Thus, it is based upon one of the four foundations of physics: the electromagnetic interaction whose synthetic mathematical formulation was made by the Scottish physicist J. C. Maxwell who published in 1873 two heavy volumes entitled *A Treatise on Electricity and Magnetism*. The discovery of the electron by J.J. Thomson in 1897 and the formulation of the theory of the atom at the beginning of the twentieth century have contributed to the first development of plasma physics. It was in 1928 that the name *plasma* was proposed for the first time by I. Langmuir, referring to blood plasma in which one finds a variety of corpuscles in movement. Experimental studies of plasmas first focused essentially on the phenomenon of electrical discharge in gas at low pressure with, for example, the formation of an electric arc. These studies initiated during the second half of the twentieth century were extended to problems related to the reflection and transmission of radio waves in the Earth's upper atmosphere (this was how the first transatlantic link was established by Marconi in 1901), which led to the discovery of the ionosphere, an atmospheric layer beyond 60 km altitude with a thickness of several hundred kilometers. As explained by the astronomer S. Chapman (1931), the ionosphere consists of gas partially ionized by solar ultraviolet radiation; therefore, it is the presence of

3

ionospheric plasma which explains why low-frequency waves can be reflected or absorbed depending on the frequency used.

With the beginning of the space age in the 1950 and 1960s, our understanding of the Earth's environment and also of the Universe significantly increased: this auspicious period saw the creation of space agencies such as the National Aeronautics and Space Administration (NASA) in 1958, the Centre National d'Études Spatiales (CNES) in 1961, and then the European Space Agency (ESA) in 1975. For example, the internal structure of the magnetosphere with the Van Allen radiation belts was discovered in 1958 by the NASA Explorer 1 probe, whereas the solar wind predicted theoretically by the American astrophysicist E. Parker[1] in 1958 (Parker, 1958) was explored for the first time in 1960 by the Russian mission Luna 2. Since these first steps in space, many other space missions have been launched to study astrophysical plasmas, such as those devoted to the Sun with the Solar & Heliospheric Observatory (SoHO) jointly launched by the ESA and NASA in 1995 and located in the vicinity of the Earth–Sun (Lagrangian) L1 point (the position in the space where the gravitational fields of the Sun and the Earth exactly balance the centrifugal force due to the rotational movement); it is to this day one of the most important solar missions to have been carried out in light of the harvest of results and its longevity. Its successor – the Solar Dynamics Observatory (SDO) – launched in 2010 currently gives the best available images of the solar corona as shown in Figure 1.1. The next space mission – Solar Orbiter – which is mainly an ESA mission, is scheduled for launch in 2018. The probe will follow an inclined orbit to study the polar regions of the Sun and will pass to its perigee at a distance of only about 45 solar radii in order to analyze, in particular, the early development of solar wind turbulence. The systematic exploration of space plasmas has allowed the investigation of many previously unknown physical phenomena, so we now distinguish this area of natural plasmas from laboratory plasmas. A distinction is also made between space plasmas accessible by *in situ* measurements and astrophysical plasmas, which are by definition more distant. In this case, the model universally chosen by astrophysicists is that of magnetohydrodynamics proposed in 1942 by the Swedish astrophysicist H. Alfvén (Alfvén, 1942). This model combines the equations of classical electrodynamics (Maxwell's equations) with fluid mechanics and describes the plasma behavior at the largest scales as a conducting mono-fluid.

Following the detonation of the first hydrogen bomb in 1952, a new area of plasma physics emerged: the so-called thermonuclear plasma physics in the context of production of energy by controlled thermonuclear fusion. When the research studies on this subject were declassified (1958), many theoretical advances had been made: the systematic mathematical treatment of plasmas

---

[1] The article submitted by E. Parker to *The Astrophysical Journal* was initially rejected by the two referees, who considered this solar wind model scientifically irrelevant. It was the editor – S. Chandrasekhar – who decided to ignore these opinions and published the paper.

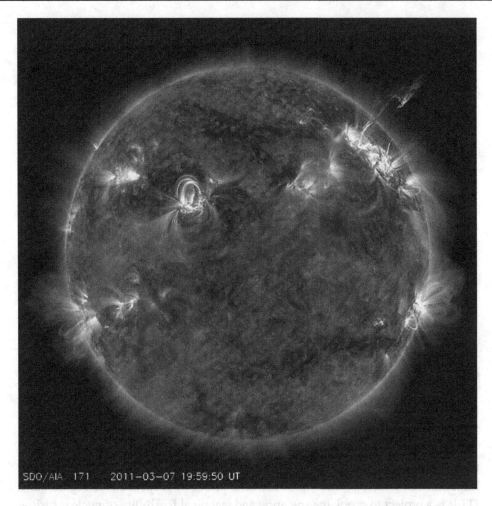

SDO/AIA 171   2011-03-07 19:59:50 UT

Figure 1.1 Solar corona observed at the wavelength 17.1 nm (Fe IX line). The solar plasma is at a temperature of about a million degrees. Image obtained by the AIA (SDO/NASA) imager (March 7, 2011); the spatial resolution is approximately 700 km; courtesy of NASA/SDO and the AIA science team.

completes the initial works of A. A. Vlasov (1938) on the kinetic equations, L. D. Landau (1946) on the damping of longitudinal space-charge waves, and H. Alfvén on magnetohydrodynamics. A central issue in this domain concerns the confinement of a hot plasma by a strong magnetic field; also many works have been devoted to the study of magnetohydrodynamic instabilities. Experimentally, the best-known magnetic confinement research technology is that of tokamaks, which were invented in the early 1950s by the Russians I. Tamm and A. Sakharov. A tokamak is a torus containing hydrogen (deuterium and tritium) that has been fully ionized and magnetized. Thus far, about 100 tokamaks have been built. The diversity of these experiences has led to a better physical understanding of

Figure 1.2 View of ITER in Cadarache. The comparison with a human (bottom right) gives an idea of the size of the tokamak. Credit: ITER Organization.

thermonuclear plasmas. It is on this basis that the International Thermonuclear Experimental Reactor (ITER) in Cadarache has been proposed (see Figure 1.2). ITER is a project to check the scientific and technical feasibility of nuclear fusion as a new source of energy for humanity; it should enter operation around the years 2025–2030. The societal issue is huge, as is its cost estimated at 20 billion euros, making it the second most expensive scientific project (after the International Space Station) ever built.

On a more modest (spatial and financial) scale, plasma technology has become invasive in our life insofar as the industrial applications are more and more numerous. Without going into detail we can cite, for example, the manufacture of small electronic components (microprocessors) in plasma reactors, or even the development of ion-propulsion engines (thrusters), which are particularly interesting for the propulsion of space probes because of the low fuel consumption (see Figure 1.3): the lunar probe SMART-1 launched in 2003 was the first ESA space mission based on the use of plasma propulsion. Eventually, the power of plasma thrusters should allow the transfer into orbit and control of the trajectory of future telecommunications satellites and should also allow control of the trajectory of planetary exploration missions.

Figure 1.3 SMART-1 (Small Missions for Advanced Research in Technology) was an ESA mission (launched in 2003 and deliberately crashed into the Moon's surface in 2006) which orbited around the Moon. It was propelled by a Hall-effect thruster (visible on this image at the bottom of the satellite). Image: ESA/SMART-1.

1.2   What Is a Plasma?

In its natural state a gas is an electrical insulator. This is because it contains no free charged particles but only neutral atoms or molecules. However, if one applies, for example, a sufficiently strong electric field, it becomes conductive: the complex phenomena that occur are called gas discharges and are due to the appearance of electrons and free ions. In such a situation, the ionized gas is characterized in general by neutrality on the macroscopic scale for which $n_e = n_i$, where n_e and n_i are the electron and ion charge densities, respectively. This neutrality is a consequence of the very intense electrostatic forces that appear whenever $n_e \neq n_i$. To measure the state of ionization of a gas, one defines the degree of ionization α as the ratio

$$\alpha \equiv \frac{n}{n_0 + n}, \tag{1.1}$$

where $n = n_e = n_i$ and n_0 is the density of neutral (i.e. non-ionized) species. This parameter allows us to distinguish weakly ionized gases, for which, typically, $10^{-10} < \alpha < 10^{-4}$, from strongly ionized gases, where $10^{-4} < \alpha < 1$. In the first category, we essentially find industrial plasmas, whereas in the second we find astrophysical and thermonuclear plasmas. It can be surprising to speak about highly ionized gas when the degree of ionization is lower than 1%. The reason is that the potential of the Coulomb interaction between two charges has a very long range (it decays as $1/r$) so that the corresponding effective section of interaction is several orders of magnitude larger than the effective section of electron–neutral-species interaction. On the other hand, weakly ionized gases are characterized by a frequency of collisions between electrons and neutral species that is higher than those of collisions between electrons and of collisions between electrons and ions.

Having defined what a plasma is, one may wonder why an initial state consisting of positive and negative charges undergoing Coulomb interactions does not naturally lead to a simple recombination of electrons with ions and thus to the disappearance of the plasma state. In fact, the maintenance of the plasma state is due to the presence, or even the combination, of several effects such as the microscopic disorder, i.e. the thermal agitation, the low reactivity of species, or the occurrence of sources of ionization such as the radiation which is particularly important in astrophysics. In general, these effects largely counterbalance the recombination induced by the Coulomb interactions. At thermodynamic equilibrium, ionization processes, are counterbalanced by recombination processes, which leads to a thermal ionization equilibrium. For example, for a gas at sufficiently high temperatures ($T \simeq 10^4$ K, i.e. an energy $(3/2)k_B T \simeq 1$ eV), there may be ionization in a collision; at even higher temperatures atoms can be ionized several times. In many cases, the gas is not in thermodynamic equilibrium and the temperatures of electrons and ions are not the same.

The name plasma is used to describe all (partially or totally) ionized gases. In summary, we have the three following families.

- Weakly ionized gases: these are plasmas in which some ions and electrons move in a sea of neutral atoms and/or molecules. In this case, it is the binary collisions between an electron (or ion) and a neutral species that determine the dynamics of charged particles. From a theoretical point of view, these plasmas are described by the Boltzmann kinetic equation.
- Strongly ionized gases with interactions between particles: an electron can be considered to be in interaction with a large number of other charged particles (due to the long range of the Coulomb force). It is the distant cumulative collisions at low deflection that determine, among other things, the plasma dynamics. These plasmas are described by a kinetic equation of the Fokker–Planck type.
- Strongly ionized gases without interaction between particles: these are diluted plasmas in which charged particles do not suffer collision and evolve only under the effect of collective electromagnetic fields due to space-charges created by all the other charges. These plasmas are described by the Vlasov kinetic equation, which is often considered the fundamental equation of plasma physics.

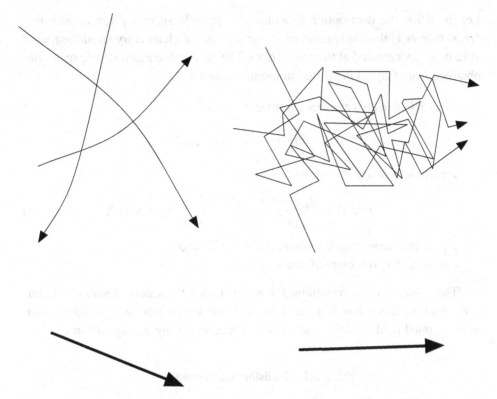

Figure 1.4 Velocities in a collisionless plasma (left) and in a collisional plasma (right), at the kinetic scale (top) and fluid scale (bottom).

Regardless of the type of plasma – collisional or collisionless – we can adopt a macroscopic (fluid) description. It is often believed that collisions are essential in order to speak about fluid models. It is even thought that a Maxwellian distribution (see Section 1.3), which is the consequence of collisions, is necessary to understand the behavior of the fluids and that it is a hidden hypothesis behind any fluid modeling. That is not true: the moment equations from which all the fluid equations derive, except the closure equation, are fully general and independent of the existence of collisions (Belmont *et al.*, 2013). However, the physical interpretation differs depending on the presence or not of collisions. In a collisional plasma, particles collide frequently and the average velocity of each particle in a given volume is equal to the local average (in this given volume) of the particle velocities. On the other hand, in a collisionless plasma each particle has an almost straight trajectory whose curvature is due to the collective field, and the average velocity can be very different from the individual velocity (see Figure 1.4).

1.3 Kinetic Description

The objective of this section is to give a rapid overview of the microscopic description of a plasma. This description is essentially based on statistical physics.

Let us define the distribution function of a particle species $f(\mathbf{r}, \mathbf{v}, t)$ such that $f(\mathbf{r}, \mathbf{v}, t) d\mathbf{r}\, d\mathbf{v}$ is the likely number of particles in the elementary six-dimensional volume $d\mathbf{r}\, d\mathbf{v}$ centered at the point $(\mathbf{r}, \mathbf{v})$. The observable macroscopic quantities obtained from f by taking different moments are the

- particle density (moment of order 0),

$$n(\mathbf{r}, t) = \int_{\mathbf{R}^3} f(\mathbf{r}, \mathbf{v}, t) d\mathbf{v}\,; \tag{1.2}$$

- particle velocity (moment of order 1),

$$\mathbf{u}(\mathbf{r}, t) = \frac{\int_{\mathbf{R}^3} \mathbf{v} f(\mathbf{r}, \mathbf{v}, t) d\mathbf{v}}{\int_{\mathbf{R}^3} f(\mathbf{r}, \mathbf{v}, t) d\mathbf{v}} = \frac{1}{n} \int_{\mathbf{R}^3} \mathbf{v} f(\mathbf{r}, \mathbf{v}, t) d\mathbf{v}\,; \tag{1.3}$$

- pressure, temperature (moment of order 2); and
- heating flux (moment of order 3).

The equation for determining $f(\mathbf{r}, \mathbf{v}, t)$ is called the kinetic equation. In general, to derive it one has to make a distinction between macroscopic forces such as the applied fields and the microscopic forces appearing during collisions.

1.3.1 Collisionless Plasma

Let us consider a volume in the space $(d\mathbf{r}, d\mathbf{v})$: the number of particles in this volume changes over time and the rate of change is given by the flux of particles across the surface of this given volume. In $\mathbf{r}$-space, this flow is

$$\int_{\mathbf{R}^5} f(\mathbf{r}, \mathbf{v}, t) \mathbf{v} \cdot \mathbf{n_r}\, dS_r\, d\mathbf{v}\,, \tag{1.4}$$

where dS_r is the surface which delimits the volume element in the $\mathbf{r}$-space, and $\mathbf{n_r}$ is the unit vector ($|\mathbf{n_r}| = 1$) orthogonal to dS_r. Similarly, in the $\mathbf{v}$-space the particle flux is

$$\int_{\mathbf{R}^5} f(\mathbf{r}, \mathbf{v}, t) \mathbf{a} \cdot \mathbf{n_v}\, dS_v\, d\mathbf{r}\,, \tag{1.5}$$

where $\mathbf{a}$ is the acceleration experienced by the particle at the level of the surface element dS_v which is oriented according to the unit vector $\mathbf{n_v}$. We obtain

$$\frac{\partial}{\partial t} \int_{\mathbf{R}^6} f(\mathbf{r}, \mathbf{v}, t) d\mathbf{r}\, d\mathbf{v} = - \int_{\mathbf{R}^5} f(\mathbf{r}, \mathbf{v}, t) \mathbf{v} \cdot \mathbf{n_r}\, dS_r\, d\mathbf{v}$$

$$- \int_{\mathbf{R}^5} f(\mathbf{r}, \mathbf{v}, t) \mathbf{a} \cdot \mathbf{n_v}\, dS_v\, d\mathbf{r}\,, \tag{1.6}$$

which gives, using the (generalized) Ostrogradsky relation,

$$\int_{\mathbf{R}^6} \left[\frac{\partial f}{\partial t} + \frac{\partial}{\partial \mathbf{r}} \cdot (\mathbf{v} f) + \frac{\partial}{\partial \mathbf{v}} \cdot (\mathbf{a} f) \right] d\mathbf{r}\, d\mathbf{v} = 0\,. \tag{1.7}$$

By definition, we write

$$\frac{\partial}{\partial \boldsymbol{\xi}} \cdot \mathbf{A} = \frac{\partial A_x}{\partial \xi_x} + \frac{\partial A_y}{\partial \xi_y} + \frac{\partial A_z}{\partial \xi_z}, \tag{1.8}$$

with the elementary volume

$$d\boldsymbol{\xi} = d\xi_x \, d\xi_y \, d\xi_z. \tag{1.9}$$

As the volume considered is arbitrary, and $\mathbf{r}$, $\mathbf{v}$, and t are independent variables, we obtain

$$\boxed{\frac{\partial f}{\partial t} + \mathbf{v} \cdot \frac{\partial f}{\partial \mathbf{r}} + \frac{\mathbf{F}}{m} \cdot \frac{\partial f}{\partial \mathbf{v}} = 0,} \tag{1.10}$$

where $\mathbf{a} = \mathbf{F}/m$. In this expression, the only force that depends on the velocity is assumed to be the Lorentz force, $q\mathbf{v} \times \mathbf{B}$; thus we can extract $\mathbf{a}$ from the derivative term (because of the cross product). Equation (1.10) is the **Vlasov equation**. We can also interpret this equation in terms of a Lagrangian and introduce the particle density conservation in a volume of the phase space which is deformed during the dynamical evolution of the plasma.

The Vlasov equation is often used to study the development of instabilities. The most remarkable study in this area is that of L. D. Landau (1946), who, from a linear analysis of the Vlasov equation and using the concept of a contour integral around singularities in complex space, was able to calculate the damping of an electrostatic Langmuir wave (see Figure 1.5). More recently, the mathematical analysis of this problem (Mouhot and Villani, 2010) and the rigorous demonstration of the Landau damping[2] in the limit of weak nonlinearities earned the French mathematician C. Villani the 2010 Fields Medal.

1.3.2 Plasma with Collisions

The effects of microscopic inter-particle forces with rapid fluctuations acting on the particles cannot be included in the force term of the Vlasov equation and must be treated separately. This leads to a new equation whose expression is

$$\frac{\partial f}{\partial t} + \mathbf{v} \cdot \frac{\partial f}{\partial \mathbf{r}} + \frac{\mathbf{F}}{m} \cdot \frac{\partial f}{\partial \mathbf{v}} = \left(\frac{\partial f}{\partial t}\right)_{c}, \tag{1.11}$$

where $(\partial f / \partial t)_c$ involves the collisions between all species of particle. Its form depends on the ionized gas considered. We see in this simple example that collisions lead simply to a modification of the Vlasov equation, which therefore retains a central position. Finally, notice that one of the roles of collisions in plasmas is to

---

[2] Note that, unlike in viscous fluid models, where the damping is an irreversible process, the Landau damping is reversible.

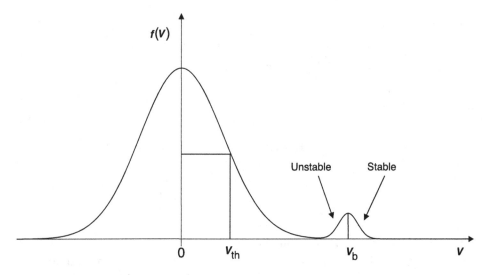

Figure 1.5 Example of distribution $f(v)$ with a beam centered on speed v_b higher than the thermal speed v_{th}. In the case of a collisionless plasma, a small perturbation – a wave – to this configuration leads to beam instability if the phase velocity of the wave v_ϕ is close to v_b with $v_\phi < v_b$, whereas there is a (Landau) damping of the wave if $v_\phi > v_b$.

quickly bring any distribution function to a quasi-Maxwellian. This demonstration was made by the Austrian physicist L. Boltzmann (1872).

1.3.3 Non-linearities

In general, the average electric field due to the charged particles can be obtained from the Poisson equation (see Chapter 2 for the notation):

$$\nabla \cdot \mathbf{E} = \frac{\rho_c}{\varepsilon_0} = \sum_s \frac{q_s}{\varepsilon_0} \int_{\mathbf{R}^3} f_s(\mathbf{r}, \mathbf{v}, t) d\mathbf{v} = \sum_s \frac{n_s q_s}{\varepsilon_0} . \tag{1.12}$$

The electric field $\mathbf{E}$ obtained (for example, after a Fourier transform) is a self-consistent field because it depends on the function f and its evaluation requires a solution of the Maxwell equations

$$\nabla \times \mathbf{E} = -\frac{\partial \mathbf{B}}{\partial t} , \tag{1.13}$$

$$\nabla \times \mathbf{B} = \mu_0 \mathbf{j} + \varepsilon_0 \mu_0 \frac{\partial \mathbf{E}}{\partial t} , \tag{1.14}$$

$$\nabla \cdot \mathbf{B} = 0 , \tag{1.15}$$

$$\mathbf{j} = \sum_s q_s \int_{\mathbf{R}^3} \mathbf{v} f_s(\mathbf{r}, \mathbf{v}, t) d\mathbf{v} , \tag{1.16}$$

with $\mathbf{j}$ the electric current and $\mathbf{B}$ the magnetic field. To solve the Vlasov equation, one has to introduce the expression for the electric field which depends on f

into expression (1.10). The non-linear nature of the Vlasov equation then appears clearly since the force involves the electric field which depends on the distribution function. In practice, this non-linear problem is essentially solved with computers.

1.4 Time Scales and Length Scales

1.4.1 Plasma Oscillations

Let us consider a plasma composed of electrons and motionless ions with the same density. At a given time we perturb this system by compressing the electrons slightly. What happens? Does the perturbation propagate like in a neutral gas via an acoustic wave, do we have a wave damping, or does the plasma oscillate without propagation? This problem is intimately linked to the notion of space-charge that is specific to ionized gases: a plasma tends to maintain itself in a state of approximate neutrality because otherwise it there quickly appear huge electric fields. For example, a deviation of 1% from neutrality in a sphere of radius r gives

$$\Delta E = \frac{1}{4\pi\varepsilon_0} \frac{e}{r^2} \Delta N = \frac{1}{4\pi\varepsilon_0} \frac{e}{r^2} \frac{1}{100} \frac{4}{3}\pi r^3 n. \tag{1.17}$$

By taking $r = 1$ cm, $n = 10^{17}$ electrons/m^3 and $e \simeq +1.6 \times 10^{-19}$ C, one gets $\Delta E = 6 \times 10^4$ V/m. Such a field imparts to an electron the acceleration $eE/m_e = 10^{16}$ m/s^2, which means that the lack of neutrality disappears very quickly.

The electrons' motion to restore neutrality will result in oscillations of charges in plasma. Let us consider a small local variation around the equilibrium state for electrons, whereas ions are assumed fixed ($m_i \gg m_e$): $n_i = n_0$ and $n_e = n_0 + n_1(\mathbf{r}, t)$ with $n_1 \ll n_0$. We have

$$\frac{\partial n_e}{\partial t} + \nabla \cdot (n_e \mathbf{u}) = 0, \tag{1.18}$$

hence, at first order

$$\frac{\partial n_1}{\partial t} + n_0 \nabla \cdot \mathbf{u} = 0. \tag{1.19}$$

Newton's second law gives us (at first order)

$$m_e \frac{\partial \mathbf{u}}{\partial t} = -e\mathbf{E}, \tag{1.20}$$

with

$$\nabla \cdot \mathbf{E} = \frac{\rho_c}{\varepsilon_0} = \frac{-n_1 e}{\varepsilon_0}. \tag{1.21}$$

The combination of these equations gives eventually

$$\frac{\partial^2 n_1}{\partial t^2} + \left(\frac{n_0 e^2}{m_e \varepsilon_0}\right) n_1 = 0, \tag{1.22}$$

whose solution is an oscillation at the angular frequency ω_{pe}:

$$\omega_{pe} = \sqrt{\frac{n_0 e^2}{m_e \varepsilon_0}}. \tag{1.23}$$

This is the so-called electronic **plasma angular frequency** (or simply plasma frequency). Any field applied at a frequency lower than the plasma frequency cannot penetrate the plasma because it reacts faster and neutralizes this field by collective (Coulomb) effects. The wave associated with the perturbation is called the Langmuir wave; it is a wave whose group velocity (the velocity at which wave energy propagates) is zero.

We have seen that the response time of electrons to a deviation from neutrality is of the order of $1/\omega_{pe}$. The phenomena that will interest us in the magneto-hydrodynamics approximation occur on large spatial and temporal scales. Therefore, we can systematically neglect the space-charge and take $\rho_c = 0$. In Table 1.1, we give the typical characteristics of several plasmas.

1.4.2   Electric Screening

The dimensional analysis of Eqs (1.20) and (1.21) gives

$$m_e u \, \omega_{pe} \sim eE \tag{1.24}$$

and

$$\frac{E}{\ell} \sim \frac{n_1 e}{\varepsilon_0}, \tag{1.25}$$

hence the characteristic scale is

$$\ell \sim \frac{n_0}{n_1} \frac{u}{\omega_{pe}}. \tag{1.26}$$

Table 1.1 Some typical plasma parameters

	n_e (m^{-3})	T_e (K)	ω_{pe} (Hz)	B (T)
Plasma reactor	10^{18}	10^4	10^{10}	10^{-2}
Cylindrical confinement	10^{22}	10^8	10^{13}	5
Tokamak	10^{20}	10^8	10^{12}	5
Earth's ionosphere	10^9–10^{12}	10^2–10^3	10^7–10^8	10^{-5}
Earth's magnetosphere	10^7	10^7	10^5	10^{-8}
Solar wind (1 UA)	10^7	10^5	10^5	10^{-9}
Solar corona	10^{14}	10^6	10^8	10^{-4}
Solar photosphere	10^{23}	10^4	10^{13}	10^{-2}
Interstellar cloud	10^8–10^9	10–100	10^6	10^{-8}

We can estimate $n_1/n_0 \sim u/V_{\text{the}}$ (for a hot plasma) with $V_{\text{the}}^2 \equiv k_B T_e/m_e$ the thermal speed of electrons. It yields

$$\ell \sim \frac{V_{\text{the}}}{\omega_{\text{pe}}} \sim \left(\frac{k_B T_e}{m_e}\right)^{1/2} \frac{1}{\omega_{\text{pe}}} \sim \left(\frac{\varepsilon_0 k_B T_e}{n_0 e^2}\right)^{1/2}; \tag{1.27}$$

this allows us to define the **Debye length** λ_D:

$$\boxed{\lambda_D \equiv \left(\frac{\varepsilon_0 k_B T_e}{n_0 e^2}\right)^{1/2}.} \tag{1.28}$$

Beyond this length scale, there is a collective effect of electric screening in the sense that the Coulomb interaction between two particles is no longer felt. If there are many particles ($N_D \gg 1$) in the Debye sphere (of radius λ_D) then the correlations between particles are weak and the collective effects (mean field) dominate.

1.4.3   Magnetic Screening

A second characteristic length scale can be introduced by considering the penetration of an electromagnetic wave in a plasma. We will consider only the motions of electrons along the y direction (see Figure 1.6). Maxwell's equations reduce to

$$\frac{\partial B_z}{\partial t} = -\frac{\partial E_y}{\partial x}, \tag{1.29}$$

$$\frac{\partial B_z}{\partial x} = \mu_0 n_0 e v_e, \tag{1.30}$$

whereas Newton's second law for electrons can be written as

$$m_e \frac{dv_e}{dt} = -eE_y. \tag{1.31}$$

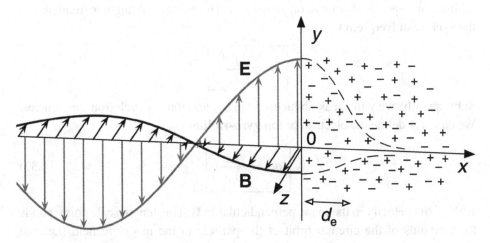

Figure 1.6 Magnetic screening: an electromagnetic wave penetrating into a plasma is damped beyond a typical length scale d_e.

The solution of this system gives

$$\frac{\partial^2 E_y}{\partial x^2} = \frac{\mu_0 n_0 e^2}{m_e} E_y = \frac{\omega_{pe}^2}{c^2} E_y \,. \tag{1.32}$$

We define the electron **inertial length** (also called electron skin depth)

$$\boxed{d_e \equiv \frac{c}{\omega_{pe}}\,,} \tag{1.33}$$

which allows us to write the physical solution of the problem:

$$E(x > 0, t) = E_0(t)\exp(-x/d_e)\,, \tag{1.34}$$

$$B(x > 0, t) = \frac{1}{d_e} \int E_0(t) dt \exp(-x/d_e)\,. \tag{1.35}$$

The electromagnetic wave is therefore damped exponentially over a characteristic length scale d_e: we refer to magnetic screening (see Figure 1.6). Physically, the application of a magnetic field in a plasma generates a collective motion of electrons and thus an electric current. The current is the source of an induced magnetic field which opposes the originally applied field. We recover Lenz's law: the effect opposes the cause that gave birth to it.

1.4.4   Cyclotron frequency

A particle of mass m and charge q moving in a magnetized medium is subjected to the Lorentz force $q\mathbf{v} \times \mathbf{B}$. Under this force and with a static magnetic field, the charge velocity can be decomposed into a parallel (to the magnetic field) component and a perpendicular component corresponding to a uniform rotating motion: one speaks about cyclotron motion. The associated angular frequency is the **cyclotron frequency**:

$$\boxed{\omega_c = \frac{|q|B}{m}\,.} \tag{1.36}$$

Subsequently, we will speak about electron ω_{ce} and ion ω_{ci} cyclotron frequencies. We can also define the electron or ion **gyro-radius**:

$$\boxed{\rho_{e,i} = \frac{v_\perp}{\omega_{ce,i}}\,,} \tag{1.37}$$

with $v_\perp$ the velocity in the plane perpendicular to $\mathbf{B}$. This length scale corresponds to the radius of the circular orbit of the particle in the magnetic field (i.e. the projection of the helicoidal motion in the plane perpendicular to the magnetic field).

1.5 From Kinetic to Fluids

1.5.1 Multi-fluid Equations

Kinetic equations describe plasmas via the distribution function $f(\mathbf{r}, \mathbf{v}, t)$, which has seven variables. The transition to a macroscopic description consists in calculating the moments of order n of f, for example the

- density,

$$n(\mathbf{r}, t) = \int_{\mathbf{R}^3} f(\mathbf{r}, \mathbf{v}, t) d\mathbf{v}; \qquad (1.38)$$

- velocity,

$$\mathbf{u}(\mathbf{r}, t) = \langle \mathbf{v} \rangle = \frac{1}{n} \int_{\mathbf{R}^3} \mathbf{v} f(\mathbf{r}, \mathbf{v}, t) d\mathbf{v}; \qquad (1.39)$$

- pressure (tensor of order 2),

$$\mathbf{P}(\mathbf{r}, t) = m \int_{\mathbf{R}^3} (\mathbf{v} - \mathbf{u})(\mathbf{v} - \mathbf{u}) f(\mathbf{r}, \mathbf{v}, t) d\mathbf{v}; \qquad (1.40)$$

- heating flux (tensor of order 3),

$$\mathbf{Q}(\mathbf{r}, t) = \frac{1}{2} m \int_{\mathbf{R}^3} (\mathbf{v} - \mathbf{u})(\mathbf{v} - \mathbf{u})(\mathbf{v} - \mathbf{u}) f(\mathbf{r}, \mathbf{v}, t) d\mathbf{v}. \qquad (1.41)$$

To establish the multi-fluid equations, we start from the collisional kinetic Eq. (1.11) that we integrate over $\mathbf{v}$. For the zeroth-order moment, we obtain the following mass continuity equation (for each particle species):

$$\frac{\partial mn}{\partial t} + \nabla \cdot (mn\mathbf{u}) = 0. \qquad (1.42)$$

For the moment of order 1, we obtain the equation of motion (for each particles species)

$$nm \left(\frac{\partial \mathbf{u}}{\partial t} + \mathbf{u} \cdot \nabla \mathbf{u} \right) = -\nabla \cdot \mathbf{P} + n\mathbf{F} + \mathbf{R}, \qquad (1.43)$$

where $\mathbf{R}$ comes from the collisional term and manifests the exchange of momentum with other particle species of the plasma. For the force, we have for example

$$\mathbf{F} = q(\mathbf{E} + \mathbf{u} \times \mathbf{B}). \qquad (1.44)$$

And so forth for the higher-order moments. Note that by this procedure the equation for the zeroth-order moment introduces the moment of order 1, the equation for the first-order moment introduces the moment of order 2, etc., hence the need to introduce a closure assumption. Among the many works dedicated to this subject, we may cite for example those based on perturbative developments of the distribution function around a Maxwellian that enable one to finally obtain the expression for the transport coefficients (Braginskii, 1965).

1.5.2   Mono-fluid Equations

If one takes the sum of all moment equations of the same order for different particle species, by introducing

- the volumic mass density,

$$\rho = \sum_s n_s m_s \,, \tag{1.45}$$

- the fluid density,

$$\mathbf{u} = \frac{\sum_s n_s m_s \mathbf{u}_s}{\sum_s n_s m_s} \,, \tag{1.46}$$

one can derive the mono-fluid equations of standard magnetohydrodynamics (see Chapter 2). In the case of a scalar[3] pressure P, one gets the inviscid (i.e. zero-viscosity) equations

$$\frac{\partial \rho}{\partial t} + \nabla \cdot (\rho \mathbf{u}) = 0 \,, \tag{1.47}$$

$$\rho \left(\frac{\partial \mathbf{u}}{\partial t} + \mathbf{u} \cdot \nabla \mathbf{u} \right) = -\nabla P + \mathbf{j} \times \mathbf{B} \,. \tag{1.48}$$

One needs to add Maxwell's equations with, e.g. the standard form of Ohm's law,

$$\mathbf{j} = \sigma (\mathbf{E} + \mathbf{u} \times \mathbf{B}) \,, \tag{1.49}$$

where σ is the electric conductivity. We can decide to stop at this order and add, for example, an adiabatic closure equation:

$$\frac{d(P\rho^{-\gamma})}{dt} = 0 \,, \tag{1.50}$$

where γ is the polytropic index. This rapid overview gives us an idea about how the fluid equations emerge from the kinetics. In this approach, precautions have to be taken with the collisional term. We can notice that this approach allows us to pass through an intermediate step prior to the equations of magnetohydrodynamics: these are the bi-fluid equations that are sometimes useful when one wants to treat the ions and electrons separately.

---

[3] The assumption of a scalar pressure can be justified rigorously only if the distribution function is fully isotropic. Such a hypothesis is justified when the plasma is strongly collisional. In collisionless plasmas the assumption of gyrotropic pressure is better adapted (Belmont et al., 2013), but in practice a scalar pressure is often used for simplicity.

2

Magnetohydrodynamics

2.1 Introduction

Magnetohydrodynamics is a branch of physics devoted to the study of the dynamics of electrically conducting fluids in the presence of magnetic fields. Magnetohydrodynamics (MHD for short) applies to most astrophysical plasmas, some laboratory plasmas, and liquid metals (e.g. mercury, sodium, gallium).

2.1.1 Electromagnetic Induction

When a conducting fluid moves in a magnetic field, it is the origin of an electric field that produces electric currents; these modify the initial magnetic field. On the other hand, the Laplace force applied to the material along the current lines changes the motion of the fluid, tending to oppose the cause that gave birth to it (Lenz's law), Thus there appears a mutual interaction of the electromagnetic and hydrodynamic effects: the conducting fluid may affect the magnetic field lines and the magnetic field can pull the fluid. The relative importance of this interaction is characterized by a dimensionless number R_m called the magnetic Reynolds number; R_m is proportional to the conductivity of the fluid, its velocity, and the dimensions of the flow. The interaction is usually weak ($R_m \ll 1$) in liquid metals and weakly ionized gases, and strong ($R_m \gg 1$) in plasmas.

The mutual interaction described above is a well-known phenomenon called electromagnetic induction, which is illustrated in Figure 2.1. In this example, a part of a conducting loop crosses a region where a magnetic field **B** is present (top). When an external force **F** is applied to the loop, which then moves at a speed **v**, an induced current i appears because of the Lorentz force:

$$\mathbf{F}_{\text{Lorentz}} = q\,\mathbf{v} \times \mathbf{B}. \qquad (2.1)$$

In other words, each of the mobile charges q inside the conductor is subjected to the Lorentz force locally in the region where a magnetic field is present. From the

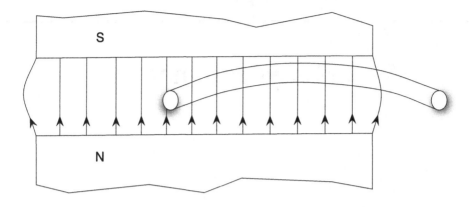

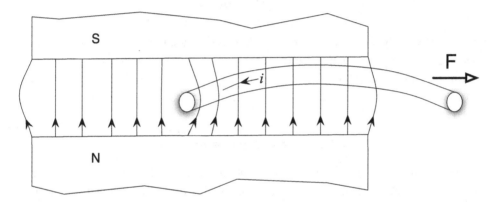

Figure 2.1 Mutual interaction of a conductor and a magnetic field: the displacement of a closed conductor in a magnetic field produces an induced electric current in the conductor. Adapted from Davidson (2001).

Biot–Savart law (1820) (quasi-stationarity is assumed here),

$$\mathbf{B}(\mathbf{r}) = \frac{\mu_0}{4\pi} \iiint \frac{\mathbf{j}(\mathbf{r}') \times (\mathbf{r} - \mathbf{r}')}{|\mathbf{r} - \mathbf{r}'|^3} \, d\mathcal{V} , \tag{2.2}$$

with $\mathbf{j}$ the electric current density, we can say that an induced magnetic field is produced in the vicinity of the loop that will modify the imposed magnetic field so that it appears as if it were being pulled by the conducting loop. Finally, the Laplace force,

$$d\mathbf{F}_{\text{Laplace}} = d\mathcal{V} \, \mathbf{j} \times \mathbf{B} , \tag{2.3}$$

will act on the loop by pushing it in a direction opposite to the initial motion. Thus, the magnetic field lines can be interpreted as immaterial threads which act on the conductor by pulling it to the left.

2.1.2   Extension to Conducting Fluids

The previous example gives us an idea of what can happen in MHD: in this case, the conducting thread is replaced by a fluid motion with a velocity **u** which depends on the position and time. The way in which the magnetic field is influenced by the velocity will depend on three parameters: the fluid velocity, its conductivity σ, and the characteristic length scale ℓ of the problem. If the fluid is not a good conductor or if its velocity is relatively small, the induced magnetic field will be negligible. Conversely, if the conductivity of the fluid or its velocity is relatively large, the induced magnetic field will be non-negligible and will significantly alter the imposed magnetic field. The length scale ℓ plays a role also in the sense that a current density in a given region can create a magnetic field whose amplitude is proportional to the size of the region (the Biot–Savart law). Therefore, it is the product $\sigma \ell u$ which will determine the relationship between the induced field and the applied field. In the limit of a perfect conductor, $\sigma \ell u \rightarrow +\infty$, the induced magnetic field and the imposed field are of the same order of magnitude. In this case, the fluid and the magnetic field move jointly: it is said that the magnetic field is "frozen" in the matter. In the opposite limit, $\sigma \ell u \rightarrow 0$, the imposed magnetic field is mainly unaffected by the conducting fluid.

2.2   Towards a Formulation of MHD

MHD began to develop in 1942 with the publication of the article "Existence of electromagnetic hydrodynamic waves" by H. Alfvén (1942) (see Figure 2.2). Up to that time, nobody was really interested in the motions of a conducting liquid in a magnetic field. In this short article (actually a half-page letter), Alfvén raises the theoretical foundations of MHD by proposing to link Maxwell's equations with those of hydrodynamics in which the Laplace force is added. He resolved this system in the incompressible and linear case and demonstrated the existence of waves which now bear his name: the **Alfvén waves**.

2.2.1   Maxwell's Equations

We will follow the same approach as Alfvén to derive the MHD equations and use a macroscopic description. This method is a powerful way to derive the MHD equations relatively quickly without entering into the complexity of the kinetic equations (see Chapter 1). Note that MHD equations also apply to liquid metals, for which it is not appropriate to introduce the plasma kinetic equations.

We start from Maxwell's equations, which are the most general expression of classical electrodynamics. They were established by J. C. Maxwell (1873) and can be written as

Figure 2.2 Hannes Alfvén (1908–1995), a Swedish astrophysicist who mainly worked on space plasmas (waves, magnetosphere, the Sun). He received the Nobel Prize in Physics in 1970 "for fundamental work and discoveries in MHD with fruitful applications in different parts of plasma physics."

$$(\text{Gauss}) \qquad \nabla \cdot \mathbf{E} = \frac{\rho_c}{\varepsilon_0}, \qquad\qquad (2.4)$$

$$\nabla \cdot \mathbf{B} = 0, \qquad\qquad (2.5)$$

$$(\text{Maxwell–Faraday}) \qquad \nabla \times \mathbf{E} = -\frac{\partial \mathbf{B}}{\partial t}, \qquad\qquad (2.6)$$

$$(\text{Maxwell–Ampère}) \qquad \nabla \times \mathbf{B} = \mu_0 \mathbf{j} + \varepsilon_0 \mu_0 \frac{\partial \mathbf{E}}{\partial t}, \qquad (2.7)$$

where ρ_c is the charge density, $\mathbf{E}$ the electric field, $\mathbf{B}$ the magnetic field, $\mathbf{j}$ the electric current density, ε_0 the permittivity of free space, and μ_0 the permeability of free space. We recall that in a non-relativistic limit the displacement current (the second term on the right in Eq. (2.7)) can be neglected. Physically, this means that the electromagnetic field is instantaneously at equilibrium and we can neglect the propagation of information by electromagnetic waves. Expression (2.7) in the non-relativistic limit implies that

$$\nabla \cdot \mathbf{j} = 0, \qquad (2.8)$$

and therefore the electric current lines are closed (like the magnetic field lines). It also means that the charge distribution is stationary on the time scales considered since the combination (2.7) and (2.4) gives

$$\frac{\partial \rho_c}{\partial t} + \nabla \cdot \mathbf{j} = 0. \qquad (2.9)$$

In the framework of non-relativistic MHD (the subject of this textbook), we will assume that the plasma is always in a state of local quasi-neutrality and thus approximately $\rho_c = 0$ (which does not mean that the electric field is zero!). One recalls that a small discrepancy from electro-neutrality implies an electronic oscillation of the plasma at the angular frequency ω_{pe} (see Chapter 1), which is much higher than the frequencies encountered in MHD.[1] We will assume for simplicity a binary plasma consisting only of electrons and protons. In these conditions, the electric current is defined as

$$\mathbf{j} = en(\mathbf{v}_i - \mathbf{v}_e), \qquad (2.10)$$

where e is the electron charge amplitude and, because of the quasi-neutrality, $n = n_e = n_i$.

2.2.2   Generalized Ohm's Law

To derive the MHD equations, we need to know the expression for the electric field. This is provided by the generalized form of Ohm's law. When a solid conductor is under the influence of an electric field, we know that the current passing through it is proportional to the electric field according to Ohm's law, $\mathbf{j} = \sigma \mathbf{E}$, in which the coefficient of proportionality is, by definition, the electric conductivity. We need now to generalize it to the case of a conducting fluid.

The idea behind MHD is that the currents at the origin of the electric fields are dragged by the matter. We know (see Figure 2.1) that a conductor moving under

---

[1] In the framework of relativistic MHD – which is not the subject of this textbook – the quasi-neutrality can be broken. Then, we often talk about Goldreich–Julian charges (Goldreich and Julian, 1969).

the influence of a magnetic field will be the origin of an electric current: in its reference frame, the conductor actually sees an electric field (by Lorentz transformation). In general, a conducting fluid always moves (with a non-relativistic speed in our case), thus we need to change Ohm's law and to add this field whose form is $\mathbf{u} \times \mathbf{B}$. The new law can be refined by adding other effects; in this textbook, we will consider only the most important corrective term, called the Hall term. This contribution becomes relevant at the ion inertial length scale, i.e. when the inertia of the ions is felt and tends to distinguish their motions locally from that of electrons (which are much more able to move quickly). The result is an electric current and hence a Laplace force mainly applied to electrons; an electric field is associated with this force. Hall MHD is used, for example, in the context of fast magnetic reconnection or high-frequency turbulence in the solar wind. Finally, the **generalized form of Ohm's law** reads

$$\mathbf{j} = \sigma \left(\mathbf{E} + \mathbf{u} \times \mathbf{B} - \frac{1}{ne} \mathbf{j} \times \mathbf{B} \right). \tag{2.11}$$

We talk about the standard form of Ohm's law when the third term on the right is negligible and the ideal form of Ohm's law when only the first two terms on the right are taken into account (i.e. when the conductivity is infinite).

2.3 Quasi-neutrality

Fluid mechanics tells us that a compressible conducting fluid is governed by the following system:

$$\frac{\partial \rho}{\partial t} + \nabla \cdot (\rho \mathbf{u}) = 0, \tag{2.12}$$

$$\rho \left(\frac{\partial \mathbf{u}}{\partial t} + \mathbf{u} \cdot \nabla \mathbf{u} \right) = -\nabla P + \rho_c \mathbf{E} + \mathbf{j} \times \mathbf{B} + \tilde{\nu} \, \Delta \mathbf{u} + \frac{\tilde{\nu}}{3} \nabla (\nabla \cdot \mathbf{u}), \tag{2.13}$$

where $\tilde{\nu}$ is the dynamic viscosity.

One recalls that, at MHD scales, there is quasi-neutrality and the Coulomb force is therefore negligible. The discrepancy from neutrality can now be evaluated by a simple dimensional analysis. If n_1 measures the deviation from neutrality n, then $\rho_c \sim e n_1$. Gauss's law (2.4) gives us the relationship

$$\frac{n_1}{n} \sim \frac{\varepsilon_0 E}{\ell e n} \sim \frac{u}{\ell c^2} \frac{B}{e n \mu_0}, \tag{2.14}$$

where c is the speed of light. (To evaluate the electric field we have used the ideal form of Ohm's law, $E \sim uB$.) The last step is to introduce a magnetic field normalized with respect to a velocity,

$$\boxed{\mathbf{b} \equiv \mathbf{B}/\sqrt{\mu_0 \rho_0}\,,} \tag{2.15}$$

and the ion inertial length $d_i \equiv c/\omega_{\mathrm{pi}}$ (see Section 1.4.4 for a definition in the context of electrons); this yields

$$\frac{n_1}{n} \sim \frac{u}{\ell c^2}\, d_i b \sim \frac{d_i}{\ell}\frac{ub}{c^2}\,. \tag{2.16}$$

In the non-relativistic case, this ratio is very small at MHD scales ($\ell \gg d_i$), but also at Hall MHD scales ($\ell < d_i$). It leads to[2]

$$\left|\frac{\rho_c \mathbf{E}}{\mathbf{j} \times \mathbf{B}}\right| \sim \frac{\rho_c E}{jB} \sim en_1 \frac{u}{j} \sim \frac{nd_i ub}{\ell c^2}\frac{u\ell \mu_0}{B} \sim \frac{u^2}{c^2}\,, \tag{2.17}$$

which demonstrates that the Coulomb force is always negligible.

2.4   Generalized (Hall) MHD Equations

The combination of Faraday's law (2.6), the generalized form of Ohm's law (2.11), and Ampère's law (2.7) gives us the following equation for the magnetic field:

$$\frac{\partial \mathbf{B}}{\partial t} = \nabla \times (\mathbf{u} \times \mathbf{B}) - \nabla \times \left(\frac{(\nabla \times \mathbf{B}) \times \mathbf{B}}{\mu_0 ne}\right) - \eta \nabla \times (\nabla \times \mathbf{B})\,, \tag{2.18}$$

where η is the **magnetic diffusivity** ($\eta\mu_0 \equiv 1/\sigma$ is the electric resistivity).

The use of vector identities (see Appendix 2) and the introduction of Ampère's law in the equation of motion allow us to get finally the non-relativistic compressible MHD equations with the Hall effect (in IS units):

$$\frac{\partial \rho}{\partial t} + \nabla \cdot (\rho \mathbf{u}) = 0\,, \tag{2.19}$$

$$\rho\left(\frac{\partial \mathbf{u}}{\partial t} + \mathbf{u} \cdot \nabla \mathbf{u}\right) = -\nabla P + \frac{1}{\mu_0}(\nabla \times \mathbf{B}) \times \mathbf{B} + \tilde{\nu}\,\Delta \mathbf{u} + \frac{\tilde{\nu}}{3}\nabla(\nabla \cdot \mathbf{u})\,, \tag{2.20}$$

$$\frac{\partial \mathbf{B}}{\partial t} = \nabla \times (\mathbf{u} \times \mathbf{B}) - \nabla \times \left(\frac{(\nabla \times \mathbf{B}) \times \mathbf{B}}{\mu_0 ne}\right) + \eta\,\Delta \mathbf{B}\,, \tag{2.21}$$

$$\nabla \cdot \mathbf{B} = 0\,. \tag{2.22}$$

We see that the fluid equations form a hierarchy of ever increasing order where each order contains a next-order quantity which must be determined from the

---

[2] Note that the calculation can be done directly without using the densities, but in that case we do not see that the density fluctuations become significant in the small-scale limit $\ell \ll d_i$.

next-order equation. Such a procedure must be closed by truncation of the hierarchy at a certain level. The common and simplest way is assuming an equation of state for the pressure which makes the energy equation useless and avoids explicitly taking into account the transport of heat. The simplest equation of state is that for an isothermal fluid (constant temperature). Then, we have $P = c_S^2 \rho$, where c_S is the speed of sound, which is constant in this case. Isothermal conditions can be applied when the temporal variations are so slow that the fluid has sufficient time to redistribute energy in order to maintain a constant heat-bath temperature. More generally, for a polytropic fluid we have $P = P_0(\rho/\rho_0)^\gamma$, where γ is the polytropic index (P_0 and ρ_0 are constants). When the time variations are so fast that no susceptible heat exchange can take place, the fluid evolves adiabatically and $\gamma = 5/3$. In these situations, the change in temperature is thus related in a simple way to the change in density (any fluid will cool during a fast expansion of the volume and heat up when the volume is compressed). One can also consider a closure at a higher order and introduce, for example, an equation for the temperature. Note that, in the incompressible case, it is not necessary to add an equation of state: indeed, the system is self-consistent since the application of the divergence operator to the equation of motion (2.20) gives us a relationship among the pressure, velocity, and magnetic fields.

The system (2.19)–(2.22) is the one that we shall study in this textbook. The Hall term is often useless at the spatial and temporal scales studied. It is therefore appropriate to evaluate the relative importance of this term over, for example, the first term on the right of Eq. (2.21). The ratio between these two non-linear terms is

$$\left| \frac{\nabla \times [(\nabla \times \mathbf{B}) \times \mathbf{B}]/(\mu_0 ne)}{\nabla \times (\mathbf{u} \times \mathbf{B})} \right| \sim \left| \frac{B/\ell}{\mu_0 ne\, \mathbf{u}} \right| \sim \frac{B}{\mu_0 ne\, \ell\, u} \sim \frac{d_i}{\ell} \frac{b}{u}, \tag{2.23}$$

where b is normalized with respect to a velocity (see expression (2.15)). If $u \sim b$, the Hall term is relevant only when the scale considered is of the order of the ion inertial length d_i. For example, in the solar wind d_i is about a few hundred kilometers (and $u \sim b$); therefore the standard MHD approximation is valid for a length scale larger than a thousand kilometers (remember that one astronomical unit (AU) is approximately 1.5×10^8 km).

Furthermore, a dimensional analysis on the ratio between the Hall term and the term on the left gives

$$\left| \frac{\nabla \times [(\nabla \times \mathbf{B}) \times \mathbf{B}]/(\mu_0 ne)}{\partial \mathbf{B}/\partial t} \right| \sim \left| \frac{B/\ell^2}{\mu_0 ne/\tau} \right| \sim \frac{\tau B}{\mu_0 ne\, \ell^2} \sim \frac{d_i}{\ell} \frac{\tau b}{\ell}, \tag{2.24}$$

where τ is the dynamical time scale considered. One can evaluate the time scale associated with the Hall term assuming that the ratio (2.24) is one; this yields

$$\tau \sim \frac{\ell}{d_i} \frac{\ell}{b} \sim \frac{\ell}{d_i} \tau_b, \tag{2.25}$$

where τ_b is the time associated with the variations of b. The time scale associated with the Hall term is therefore, in general, shorter than that for the standard MHD.

2.4.1   The Incompressible Limit

It is often interesting to consider the incompressible MHD limit. Although incompressibility does not necessarily imply that the density is uniform in space (a counter-example is given by the oceans, which are mainly incompressible with a thermal stratification that leads to spatial density variations), for simplicity we will make this assumption. For example, liquid metals are very well described by this approximation, and the density fluctuations in the solar wind are often of the order of only 5% (especially at 1 AU). Additionally, the assumption of incompressibility allows us to tackle more easily problems that are of very complex nature, such as turbulence (see Part IV).

We shall assume a constant mass density $\rho = \rho_0$ (we also have $n = n_0$). It is then very convenient to introduce a magnetic field normalized with respect to a velocity. Then, we get the system

$$\nabla \cdot \mathbf{u} = 0, \tag{2.26}$$

$$\frac{\partial \mathbf{u}}{\partial t} + \mathbf{u} \cdot \nabla \mathbf{u} = -\nabla \left(\frac{P}{\rho_0} \right) + (\nabla \times \mathbf{b}) \times \mathbf{b} + \nu \, \Delta \mathbf{u}, \tag{2.27}$$

$$\frac{\partial \mathbf{b}}{\partial t} = \nabla \times (\mathbf{u} \times \mathbf{b}) - d_i \nabla \times ((\nabla \times \mathbf{b}) \times \mathbf{b}) + \eta \, \Delta \mathbf{b}, \tag{2.28}$$

$$\nabla \cdot \mathbf{b} = 0, \tag{2.29}$$

where d_i is the ion inertial length and $\nu \equiv \tilde{\nu}/\rho_0$ is the kinematic viscosity. We immediately see the benefit of using this approach: only the scale d_i is now present. Sometimes, one prefers to write this system in a more symmetrical way. To do this, we must make use of vector identities. After a few calculations, we obtain the following incompressible MHD equations:

$$\nabla \cdot \mathbf{u} = 0, \tag{2.30}$$

$$\frac{\partial \mathbf{u}}{\partial t} + \mathbf{u} \cdot \nabla \mathbf{u} = -\nabla P_* + \mathbf{b} \cdot \nabla \mathbf{b} + \nu \, \Delta \mathbf{u}, \tag{2.31}$$

$$\frac{\partial \mathbf{b}}{\partial t} + \mathbf{u} \cdot \nabla \mathbf{b} = \mathbf{b} \cdot \nabla \mathbf{u} - d_i \nabla \times ((\nabla \times \mathbf{b}) \times \mathbf{b}) + \eta \, \Delta \mathbf{b}, \tag{2.32}$$

$$\nabla \cdot \mathbf{b} = 0, \tag{2.33}$$

where $P_* \equiv P/\rho_0 + b^2/2$ is the total pressure. It is useful to evaluate the relative importance of the non-linear terms over the dissipative terms. The kinetic R_e and magnetic R_m Reynolds numbers quantify this importance. Their definitions are

$$R_e = \frac{UL}{\nu}, \quad R_m = \frac{UL}{\eta}, \tag{2.34}$$

where U and L are the characteristic macroscopic velocity and length scale of the fluid. Note that, unlike the Navier–Stokes equations (when a uniform velocity is present), the MHD equations are not Galilean invariant in the presence of a uniform magnetic field $\mathbf{b}_0$. The consequence is that $\mathbf{b}_0$ may have a strong impact on the fluid dynamics, with the development of anisotropy (see Chapter 12) and the existence of waves (see Chapter 4).

2.4.2 Electron MHD

The system (2.31)–(2.32) allows us to better highlight the similarities to (or differences from) the purely hydrodynamic case. We see that the Hall term breaks the self-similarity of the MHD equations by introducing a characteristic length scale d_i. We will see below that the Hall term removes the degeneracy of Alfvén waves and brings out waves with right and left polarities. Actually, recent studies have shown that the Hall term leads to a spontaneous symmetry breaking in the non-linear dynamics (Meyrand and Galtier, 2012). In the case where the Hall term dominates – which means in the limit of small space and time scales – the system reduces to

$$\frac{\partial \mathbf{b}}{\partial t} = -d_i \, \nabla \times ((\nabla \times \mathbf{b}) \times \mathbf{b}) + \eta \, \Delta \mathbf{b} \,. \tag{2.35}$$

One speaks about electron MHD or simply EMHD. In this limit, the ions become too heavy to follow the electrons' motions and therefore constitute a neutralizing background on which electrons move. Then, the electric current is carried only by electrons.

2.4.3 Ideal MHD

Ideal MHD corresponds to $\eta = 0$. It is often used in astrophysics not because the conductivity σ is infinitely large but because the length scales are very large. To be more precise, let us take the example of the solar photosphere where we have in the magnetic loops of active regions the typical values

$$B = 10^{-2} \, \text{T} \,, \tag{2.36}$$

$$\ell = 10^{6} \, \text{m} \,, \tag{2.37}$$

$$u = 10^{3} \, \text{m/s} \,, \tag{2.38}$$

$$\sigma = 10^{3} \, \Omega^{-1} \, \text{m}^{-1} \,. \tag{2.39}$$

This gives

$$j \sim \frac{B}{\mu_0 \ell} = 10^{-2} \, \text{A/m}^2 \,, \tag{2.40}$$

$$|\mathbf{u} \times \mathbf{B}| \sim uB = 10 \, \text{V/m} \,, \tag{2.41}$$

hence

$$\frac{j}{\sigma} \sim 10^{-5} \, \text{V/m}, \tag{2.42}$$

which is negligible compared with $|\mathbf{u} \times \mathbf{B}|$. Therefore, it is quite relevant to consider the ideal MHD as a first approximation for this medium. Note finally that we will reserve the word inviscid (vocabulary used in fluid mechanics) for the case $\tilde{v} = 0$. Therefore, the ideal and inviscid MHD corresponds to $\tilde{v} = \eta = 0$.

2.5 Examples of Electrically Conducting Fluids

Plasmas are abundant in the Universe since they account for more than 99% of visible matter. The simplest plasma is that of hydrogen, which is made of free electrons and protons at equal densities (fully ionized gas). As a first approximation, we can consider that stars are plasma balls. In the case of the Sun, the surface is composed of 73.5% hydrogen, 25% helium, and only traces ($\sim$1%) of heavier elements such as iron, which is ionized several times (Priest, 2014). Figure 1.1 shows precisely the solar corona seen by the AIA imager on NASA's SDO probe at the wavelength 17.1 nm (the Fe IX line): the very small amount of iron (approximately 0.15% of the visible matter of the Sun) ionized several times is used to see the structures of the solar corona at about 1 million degrees. In this image there appear active regions (the brightest parts) where magnetic loops are present. The darker regions correspond to coronal holes; they often appear at the level of the poles (here at the bottom of the image) and are the sources of the turbulent solar wind (see Chapter 12) that sweeps the interplanetary medium at speeds between 300 and 900 km/s. The magnetic structures of the solar corona are very dynamic; they may become unstable and cause solar flares after a process of magnetic reconnection (see Chapter 7). Large amounts of plasma are then projected into the interplanetary medium with – as one might imagine – major risks of damage for various satellites (e.g. telecommunications satellites). MHD is generally used to describe this solar dynamics. It is also widely used for the modeling of many other astrophysical objects or processes, e.g. accretion disks, magnetic fields around stars, solar wind turbulence, and the origin and dynamics of the Earth's magnetosphere. The importance of MHD in all these astrophysical events is due to the fact that the dimensions, speeds, and conductivities of the plasma are often very large, meaning that the matter and the magnetic field are strongly coupled.

Electrically conducting fluids can also be found in planets in the form of liquid metals. It is believed that such fluids are the source of the planetary magnetic fields which are now well known to exist in our solar system thanks to several space missions. The physics devoted to the question of magnetic-field production is called *dynamo* physics and will be presented in detail in Chapter 5. It is a

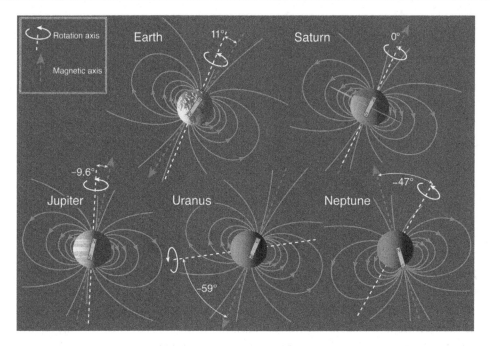

Figure 2.3 Tilts of planetary magnetic fields with respect to their rotating axes. Courtesy of F. Bagenal and S. Bartlett.

very active subject of research where several important questions still remain unanswered. In Figure 2.3 we give five examples of planetary magnetic fields. Almost systematically, a tilt of the magnetic dipolar axis with respect to the rotation axis is found. In the case of the Earth, the liquid metal is localized in the outer core – between the inner core made mainly of solid iron and the mantle. Even though our knowledge is more limited for other planets, it is believed that the internal structure and the localization of the electrically conducting fluids are probably not exactly the same as for the Earth.

Plasma laboratories are necessarily limited in size. The technologies of plasma production are numerous, but the best-known example is probably that of tokamaks (see Figure 1.2) and in particular the international ITER project. Tokamak plasmas are magnetically confined by an intense field of several teslas at a temperature of approximately 100 million degrees. These plasmas are composed of light nuclei such as deuterium and tritium to facilitate the thermonuclear fusion reaction. The study of the stability and magnetic confinement of such a plasma will be discussed in detail in Chapters 8 to 10.

At the terrestrial scale many technical applications of MHD were considered, some as early as in the nineteenth century. However, their practical development faces a major difficulty: liquid metals and ionized gases are, in general, less conductive than copper, so they can substitute for it in electrical engineering

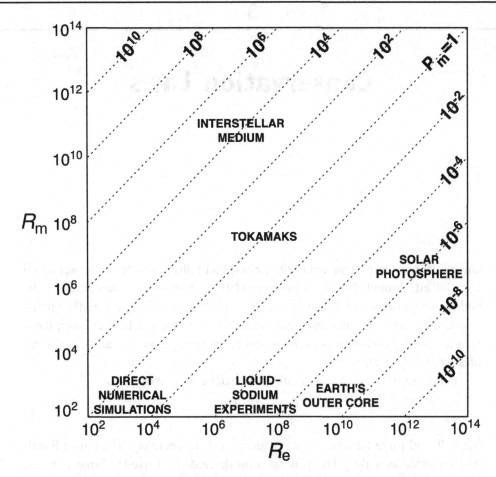

Figure 2.4 Estimates of the kinetic R_e and magnetic R_m Reynolds numbers for various media. The diagonals correspond to constant magnetic Prandtl numbers $P_m \equiv R_m/R_e = \nu/\eta$.

only in very special cases. Major industrial developments involve liquid metals (electromagnetic pumping of sodium used as a coolant in some nuclear reactors, etc.). In these applications the induced magnetic field is relatively weak since the scales and speeds are very limited in laboratories. Note that in the years 1960–1970 industrial developments around MHD propulsion in the maritime domain were considered because sea water conducts electricity, but in fact its very low conductivity ($\eta \simeq 8 \times 10^5$ m²/s whereas $\nu \simeq 8 \times 10^{-6}$ m²/s) greatly limits its use.

To conclude, typical values of the kinetic and magnetic Reynolds numbers for various plasmas and liquid metals whose description is governed (in part) by MHD are given in Figure 2.4.

3

Conservation Laws

One of the most important and useful principles of physics is that of conservation laws linked, through Noether's theorem (1918), to invariance properties of the underlying equations. In this chapter, conservation laws which govern MHD fluids (mass, momentum, energy, etc.) will be established. As will be seen later, these laws are fundamental means to characterize, for example, shocks and discontinuities (see Chapter 6).

In the general case, a conservation law can be written as

$$\frac{\partial T_{jkl...}}{\partial t} = -\frac{\partial F_{ijkl...}}{\partial x_i},\tag{3.1}$$

where **T** and **F** are tensors of rank N and $N+1$, respectively. The tensor **F** will be interpreted as a flux. The general form derived for a fixed volume $\mathcal{V}$ gives, after application of Ostrogradski's theorem (also called the divergence theorem; see Appendix 2),

$$\frac{d}{dt}\iiint T_{jkl...}\,d\mathcal{V} = -\oiint F_{ijkl...}n_i\,d\mathcal{S}\,,\tag{3.2}$$

where $\mathcal{S}$ is a closed surface delimiting the volume $\mathcal{V}$ and **n** is the vector normal to the surface. Expression (3.2) is valid regardless of the chosen (fixed) volume: for example, one can consider the plasma volume, i.e. the volume $\mathcal{V}_\infty$ at the surface of which the plasma is always at rest. This choice will allow us to highlight the roles of some MHD invariants.

3.1   Mass

The local equation of mass conservation has already been introduced in the previous chapter:

$$\frac{\partial \rho}{\partial t} = -\nabla \cdot (\rho \mathbf{u})\,,\tag{3.3}$$

where $\rho\mathbf{u}$ is the mass flux. Its integral form is

$$\frac{d}{dt} \iiint \rho \, d\mathcal{V} = -\oiint \rho \mathbf{u} \cdot \mathbf{n} \, d\mathcal{S}, \tag{3.4}$$

which shows that the mass in a given volume changes only if material crosses its surface $\mathcal{S}$. In particular, if one chooses the volume $\mathcal{V}_\infty$ for which the plasma velocity is zero at the surface, the mass is conserved.

3.2 Momentum

From Eqs. (2.20) and (3.3), it can be shown that the local equation of momentum conservation reads

$$\frac{\partial \rho \mathbf{u}}{\partial t} = -\nabla \cdot (\rho \mathbf{u}\mathbf{u}) - \nabla P + \frac{1}{\mu_0}(\nabla \times \mathbf{B}) \times \mathbf{B} + \tilde{\nu} \, \Delta \mathbf{u} + \frac{\tilde{\nu}}{3} \nabla(\nabla \cdot \mathbf{u}). \tag{3.5}$$

By noticing that

$$(\nabla \times \mathbf{B}) \times \mathbf{B} = -\nabla\left(\frac{B^2}{2}\right) + \mathbf{B} \cdot \nabla \mathbf{B}$$

$$= \nabla \cdot \left(-\frac{B^2}{2}\mathbf{I} + \mathbf{B}\mathbf{B}\right), \tag{3.6}$$

where $\mathbf{I}$ is the identity tensor, one obtains

$$\frac{\partial \rho \mathbf{u}}{\partial t} = -\nabla \cdot \left(\rho \mathbf{u}\mathbf{u} + \left(\frac{B^2}{2\mu_0}\mathbf{I} - \frac{\mathbf{B}\mathbf{B}}{\mu_0}\right) + P\mathbf{I} - \tilde{\nu}\,\Pi\right), \tag{3.7}$$

with the normalized viscosity stress tensor

$$\nabla \cdot \Pi|_j = \partial_i \Pi_{ij} = \partial_i \left(\partial_i u_j + \partial_j u_i - \frac{2}{3}\delta_{ij}\,\partial_k u_k\right). \tag{3.8}$$

We recognize in (3.7) the Reynolds stress tensor $\rho\mathbf{u}\mathbf{u}$ and the Maxwell (magnetic terms) tensor. The integral form of the momentum conservation is

$$\frac{d}{dt} \iiint \rho \mathbf{u} \, d\mathcal{V} = -\oiint \left(\rho \mathbf{u}\mathbf{u} + P\mathbf{I} + \frac{B^2}{2\mu_0}\mathbf{I} - \frac{\mathbf{B}\mathbf{B}}{\mu_0} - \tilde{\nu}\,\Pi\right) \cdot \mathbf{n} \, d\mathcal{S}. \tag{3.9}$$

The variation of momentum may have several origins: there are the contributions of the momentum flux (the first two terms on the right-hand side), the Maxwell tensor, and the viscosity stress tensor. Let us consider the volume $\mathcal{V}_\infty$. The viscosity stress tensor involves velocity derivatives which are *a priori* non-null on $\mathcal{S}_\infty$; thus, it is necessary to consider an inviscid fluid ($\tilde{\nu} = 0$) in order to remove its contribution. Then, we can trivially eliminate the momentum flux. Unlike in hydrodynamics, under these hypotheses the momentum is not conserved in MHD because the magnetic field carries momentum.

3.3 Energy

We shall write the equation of energy conservation. The first question that arises concerns the nature of this energy. We will see that we need to consider the total energy of the system, namely the sum of the kinetic, magnetic, and internal energies:[1]

$$E \equiv \frac{\rho u^2}{2} + \frac{B^2}{2\mu_0} + \rho e. \tag{3.10}$$

For our study, we will use a polytropic closure, with $P = P_0(\rho/\rho_0)^\gamma$, and thus the internal energy $\rho e = P/(\gamma - 1)$.

The evolution equation of the kinetic energy is derived in two steps. First, we note that

$$\frac{1}{2}\rho\frac{\partial u^2}{\partial t} = \rho \mathbf{u} \cdot \frac{\partial \mathbf{u}}{\partial t}$$

$$= \mathbf{u} \cdot \left(-\rho \mathbf{u} \cdot \nabla \mathbf{u} - \nabla P + \mathbf{j} \times \mathbf{B} + \tilde{\nu} \Delta \mathbf{u} + \frac{\tilde{\nu}}{3}\nabla(\nabla \cdot \mathbf{u}) \right). \tag{3.11}$$

The use of vectorial identities (see Appendix 2) and Ohm's law give the following relations:

$$\mathbf{u} \cdot (\mathbf{j} \times \mathbf{B}) = -\mathbf{j} \cdot (\mathbf{u} \times \mathbf{B}) = \mathbf{j} \cdot \left(\mathbf{E} - \frac{\mathbf{j}}{\sigma} - \frac{1}{ne}\mathbf{j} \times \mathbf{B} \right)$$

$$= \mathbf{j} \cdot \mathbf{E} - \frac{j^2}{\sigma}, \tag{3.12}$$

$$\mathbf{u} \cdot \Delta \mathbf{u} = \mathbf{u} \cdot \nabla(\nabla \cdot \mathbf{u}) - \mathbf{u} \cdot (\nabla \times \mathbf{w})$$

$$= \mathbf{u} \cdot \nabla(\nabla \cdot \mathbf{u}) + \nabla \cdot (\mathbf{u} \times \mathbf{w}) - w^2, \tag{3.13}$$

$$\mathbf{u} \cdot \nabla(\nabla \cdot \mathbf{u}) = \nabla \cdot ((\nabla \cdot \mathbf{u})\mathbf{u}) - (\nabla \cdot \mathbf{u})^2, \tag{3.14}$$

with $\mathbf{w} \equiv \nabla \times \mathbf{u}$ the **vorticity**; hence we obtain the equation

$$\frac{1}{2}\rho\frac{\partial u^2}{\partial t} = -\nabla \cdot \left(P\mathbf{u} - \frac{4}{3}\tilde{\nu}(\nabla \cdot \mathbf{u})\mathbf{u} - \tilde{\nu}\mathbf{u} \times \mathbf{w} \right)$$

$$- \rho \mathbf{u} \cdot \nabla \left(\frac{u^2}{2} \right) + P\nabla \cdot \mathbf{u} + \mathbf{j} \cdot \mathbf{E} - \frac{j^2}{\sigma} - \tilde{\nu}w^2 - \frac{4}{3}\tilde{\nu}(\nabla \cdot \mathbf{u})^2. \tag{3.15}$$

Additionally, we have

---

[1] It is easy to show that the electric energy is negligible in comparison with the magnetic one in the non-relativistic case (the energy ratio is $\propto u^2/c^2$).

$$\frac{u^2}{2}\frac{\partial\rho}{\partial t} = -\frac{u^2}{2}\nabla\cdot(\rho\mathbf{u}),\qquad(3.16)$$

which gives finally the equation for the kinetic energy:

$$\frac{1}{2}\frac{\partial\rho u^2}{\partial t} = -\nabla\cdot\left(\frac{\rho u^2}{2}\mathbf{u} + P\mathbf{u} - \frac{4}{3}\tilde{\nu}(\nabla\cdot\mathbf{u})\mathbf{u} - \tilde{\nu}\,\mathbf{u}\times\mathbf{w}\right)$$

$$+ P\nabla\cdot\mathbf{u} + \mathbf{j}\cdot\mathbf{E} - \frac{j^2}{\sigma} - \tilde{\nu}w^2 - \frac{4}{3}\tilde{\nu}(\nabla\cdot\mathbf{u})^2.\qquad(3.17)$$

The evolution equation of the magnetic energy is trivially derived if one uses Faraday's law:

$$\frac{1}{2\mu_0}\frac{\partial B^2}{\partial t} = \frac{1}{\mu_0}\mathbf{B}\cdot\frac{\partial\mathbf{B}}{\partial t} = -\frac{1}{\mu_0}\mathbf{B}\cdot(\nabla\times\mathbf{E})$$

$$= -\nabla\cdot\left(\frac{\mathbf{E}\times\mathbf{B}}{\mu_0}\right) - \mathbf{E}\cdot\left(\frac{\nabla\times\mathbf{B}}{\mu_0}\right)$$

$$= -\nabla\cdot\mathcal{P} - \mathbf{j}\cdot\mathbf{E},\qquad(3.18)$$

where $\mathcal{P}$ is the **Poynting vector**.

The evolution equation of the internal energy for a polytropic fluid is derived by writing

$$\frac{\partial\rho e}{\partial t} = \frac{1}{\gamma-1}\frac{\partial P}{\partial t} = \frac{\gamma}{\gamma-1}\frac{P}{\rho}\frac{\partial\rho}{\partial t} = -\frac{\gamma}{\gamma-1}\frac{P}{\rho}\nabla\cdot(\rho\mathbf{u})$$

$$= -\frac{\gamma}{\gamma-1}P\nabla\cdot\mathbf{u} - \frac{\gamma}{\gamma-1}\frac{P}{\rho}\mathbf{u}\cdot\nabla\rho.\qquad(3.19)$$

By noticing that

$$\nabla\cdot(\rho e\mathbf{u}) = \mathbf{u}\cdot\nabla(\rho e) + \rho e\nabla\cdot\mathbf{u} = \frac{\gamma}{\gamma-1}\frac{P}{\rho}\mathbf{u}\cdot\nabla\rho + \frac{1}{\gamma-1}P\nabla\cdot\mathbf{u},\qquad(3.20)$$

we find

$$\frac{\partial\rho e}{\partial t} = -\nabla\cdot(\rho e\mathbf{u}) - P\nabla\cdot\mathbf{u}.\qquad(3.21)$$

Equations (3.17), (3.18), and (3.21) are very interesting because they reveal information about the energy transfer. We see that the kinetic and magnetic energies exchange information via the electromagnetic power (density), $\mathbf{j}\cdot\mathbf{E}$, whereas for the kinetic and internal energies this is done via the pressure dilatation, $P\nabla\cdot\mathbf{u}$, which is by nature purely compressible.

The addition of Eqs. (3.17), (3.18), and (3.21) gives finally the local form of the total energy conservation:

$$\frac{\partial E}{\partial t} = -\nabla\cdot\left[\left(\frac{\rho u^2}{2} + P + \rho e\right)\mathbf{u} + \mathcal{P} - \frac{4}{3}\tilde{\nu}(\nabla\cdot\mathbf{u})\mathbf{u} - \tilde{\nu}\,\mathbf{u}\times\mathbf{w}\right]$$

$$- \tilde{\nu}w^2 - \mu_0\eta j^2 - \frac{4}{3}\tilde{\nu}(\nabla\cdot\mathbf{u})^2.\qquad(3.22)$$

The variation of energy is due to two types of term: flux terms and source terms (the source terms are in the second line). The sources are all negative and depend on the viscosity or magnetic diffusivity. In other words, there is an irreversible source of dissipation for the energy which is absent in the ideal and inviscid case ($\tilde{\nu} = \eta = 0$). In the divergence, we find the kinetic energy flux (the first three terms on the right-hand side), the Poynting vector, and a viscous contribution non-linearly dependent on the velocity. Thus, the magnetic contribution is localized solely in the Poynting flux. By using Ohm's law, we note that this term is null if the plasma is at rest (then there is no current). The integral form of the energy conservation in the inviscid and ideal case can be written as

$$\frac{d}{dt} \iiint E \, d\mathcal{V} = - \iint \left[\left(\frac{\rho u^2}{2} + P + \rho e \right) \mathbf{u} + \mathcal{P} \right] \cdot \mathbf{n} \, d\mathcal{S}. \tag{3.23}$$

If we consider the particular case of the volume $\mathcal{V}_\infty$, the flux vanishes on the surface $\mathcal{S}_\infty$ and the energy is conserved. The total energy is therefore an inviscid and ideal invariant of the generalized MHD.[2]

3.4 Cross-helicity

We shall demonstrate that in the particular case of standard MHD, there is a second ideal and inviscid invariant involving the velocity and the magnetic field; it is the cross-helicity H^c:

$$H^c \equiv \mathbf{u} \cdot \mathbf{B}. \tag{3.24}$$

We will see below (in Exercise I.b) that the cross-helicity measures the mutual degree of linkage of the vorticity and magnetic fields (Moffatt, 1969). We have

$$\frac{\partial H^c}{\partial t} = \mathbf{u} \cdot \frac{\partial \mathbf{B}}{\partial t} + \mathbf{B} \cdot \frac{\partial \mathbf{u}}{\partial t}, \tag{3.25}$$

with

$$\mathbf{B} \cdot \frac{\partial \mathbf{u}}{\partial t} = \mathbf{B} \cdot \left(-\mathbf{u} \cdot \nabla \mathbf{u} - \frac{1}{\rho} \nabla P + \frac{\tilde{\nu}}{\rho} \left(\Delta \mathbf{u} + \frac{1}{3} \nabla (\nabla \cdot \mathbf{u}) \right) \right), \tag{3.26}$$

$$\mathbf{u} \cdot \frac{\partial \mathbf{B}}{\partial t} = \mathbf{u} \cdot \left(\nabla \times (\mathbf{u} \times \mathbf{B}) - \nabla \times \left(\frac{\mathbf{j} \times \mathbf{B}}{ne} \right) + \eta \, \Delta \mathbf{B} \right). \tag{3.27}$$

By noticing that

$$\mathbf{B} \cdot (\mathbf{u} \cdot \nabla \mathbf{u}) = \nabla \cdot \left(\frac{u^2}{2} \mathbf{B} \right) + \mathbf{w} \cdot (\mathbf{u} \times \mathbf{B}), \tag{3.28}$$

$$\mathbf{u} \cdot (\nabla \times (\mathbf{u} \times \mathbf{B})) = -\nabla \cdot (\mathbf{u} \times (\mathbf{u} \times \mathbf{B})) + \mathbf{w} \cdot (\mathbf{u} \times \mathbf{B}), \tag{3.29}$$

---

[2] In the ideal and inviscid limit ($\tilde{\nu}, \eta \to 0^+$; but $\tilde{\nu}, \eta \neq 0$), the situation is more difficult to analyze since the terms with velocity derivatives may tend to infinity. For example, what happens then for the source terms? We will return to this fundamental issue in the last part of the book, which is devoted to turbulence.

$$\mathbf{u} \cdot \left(\nabla \times \left(\frac{\mathbf{j} \times \mathbf{B}}{ne} \right) \right) = -\nabla \cdot \left(\mathbf{u} \times \left(\frac{\mathbf{j} \times \mathbf{B}}{ne} \right) \right) + \mathbf{w} \cdot \left(\frac{\mathbf{j} \times \mathbf{B}}{ne} \right), \qquad (3.30)$$

$$\frac{1}{\rho} \mathbf{B} \cdot \nabla P = \frac{\gamma}{\gamma - 1} \nabla \cdot \left(\frac{P}{\rho} \mathbf{B} \right) = \nabla \cdot (\gamma e \mathbf{B}), \qquad (3.31)$$

we obtain

$$\frac{\partial H^c}{\partial t} = -\nabla \cdot \left[\left(\frac{u^2}{2} + \gamma e \right) \mathbf{B} + \mathbf{u} \times (\mathbf{u} \times \mathbf{B}) - \mathbf{u} \times \left(\frac{\mathbf{j} \times \mathbf{B}}{ne} \right) \right]$$
$$- \mathbf{w} \cdot \left(\frac{\mathbf{j} \times \mathbf{B}}{ne} \right) + \eta \mathbf{u} \cdot \Delta \mathbf{B} + \frac{\tilde{\nu}}{\rho} \mathbf{B} \cdot \left(\Delta \mathbf{u} + \frac{1}{3} \nabla (\nabla \cdot \mathbf{u}) \right). \quad (3.32)$$

The viscous terms can be rewritten by noting that

$$\mathbf{u} \cdot \Delta \mathbf{B} = -\mu_0 \mathbf{u} \cdot (\nabla \times \mathbf{j}) = \mu_0 \nabla \cdot (\mathbf{u} \times \mathbf{j}) - \mu_0 \mathbf{j} \cdot \mathbf{w}, \qquad (3.33)$$

$$\mathbf{B} \cdot \Delta \mathbf{u} = \mathbf{B} \cdot (\nabla (\nabla \cdot \mathbf{u}) - (\nabla \times \mathbf{w}))$$
$$= \nabla \cdot ((\nabla \cdot \mathbf{u}) \mathbf{B} + \mathbf{B} \times \mathbf{w}) - \mu_0 \mathbf{j} \cdot \mathbf{w}, \qquad (3.34)$$

$$\mathbf{B} \cdot (\nabla (\nabla \cdot \mathbf{u})) = \nabla \cdot ((\nabla \cdot \mathbf{u}) \mathbf{B}), \qquad (3.35)$$

hence the local form of the cross-helicity conservation:

$$\frac{\partial H^c}{\partial t} = -\nabla \cdot \left[\left(\frac{u^2}{2} + \gamma e \right) \mathbf{B} + \mathbf{u} \times (\mathbf{u} \times \mathbf{B}) - \mathbf{u} \times \left(\frac{\mathbf{j} \times \mathbf{B}}{ne} \right) - \mu_0 \eta \, \mathbf{u} \times \mathbf{j} \right]$$
$$+ \frac{\tilde{\nu}}{\rho} \nabla \cdot \left(\mathbf{B} \times \mathbf{w} + \frac{4}{3} (\nabla \cdot \mathbf{u}) \mathbf{B} \right) - \mathbf{w} \cdot \left(\frac{\mathbf{j} \times \mathbf{B}}{ne} \right)$$
$$- \left(\eta + \frac{\tilde{\nu}}{\rho} \right) \mu_0 \mathbf{j} \cdot \mathbf{w}. \qquad (3.36)$$

As for the energy, the variation of the cross-helicity is due to two types of term: flux terms and source terms. The main difference is that the source terms have no defined sign and may therefore behave as sources of dissipation or production whose origin is not exclusively viscous since the Hall term contributes. For this reason, the cross-helicity cannot be an ideal and inviscid invariant of Hall MHD. In the particular case of standard inviscid and ideal MHD, the integral form of the cross-helicity conservation is reduced to

$$\frac{d}{dt} \iiint H^c \, d\mathcal{V} = - \oiint \left[\left(\frac{u^2}{2} + \gamma e \right) \mathbf{B} + \mathbf{u} \times (\mathbf{u} \times \mathbf{B}) \right] \cdot \mathbf{n} \, d\mathcal{S}. \qquad (3.37)$$

If we consider the particular case of the volume $\mathcal{V}_\infty$, the flux vanishes on the surface $\mathcal{S}_\infty$ (the pressure and hence the density and internal energy are then constant) and thus the integral vanishes. The cross-helicity is thus an inviscid and ideal invariant of standard MHD.

It is relevant to note that, in incompressible standard MHD (which is equivalent to the limit $\gamma \to +\infty$), we get simply

$$\frac{\partial H^c}{\partial t} = -\nabla \cdot \left[\left(\frac{u^2}{2} + \frac{P}{\rho_0} \right) \mathbf{B} + \mathbf{u} \times (\mathbf{u} \times \mathbf{B}) - \mu_0 \eta \, \mathbf{u} \times \mathbf{j} - \nu \mathbf{B} \times \mathbf{w} \right]$$
$$- (\eta + \nu) \mu_0 \mathbf{j} \cdot \mathbf{w} . \tag{3.38}$$

In this case, these remains only a single source term, which is by nature purely viscous.

3.5 Magnetic Helicity

The generalized MHD equations have a purely magnetic invariant; it is the so-called magnetic helicity (Woltjer, 1958):

$$\boxed{H^{\mathrm{m}} \equiv \mathbf{A} \cdot \mathbf{B} ,} \tag{3.39}$$

where $\mathbf{A}$ is the vector potential ($\mathbf{B} = \nabla \times \mathbf{A}$) which is defined up to a gauge. As will be shown below (Sections 3.6 and 3.7), the magnetic helicity provides a measure of the degree of structural complexity of a magnetic field. We have

$$\frac{\partial H^{\mathrm{m}}}{\partial t} = \mathbf{A} \cdot \frac{\partial \mathbf{B}}{\partial t} + \mathbf{B} \cdot \frac{\partial \mathbf{A}}{\partial t} . \tag{3.40}$$

With Faraday's law we can write

$$\mathbf{A} \cdot \frac{\partial \mathbf{B}}{\partial t} = -\mathbf{A} \cdot (\nabla \times \mathbf{E}) = -\nabla \cdot (\mathbf{E} \times \mathbf{A}) - \mathbf{B} \cdot \mathbf{E} , \tag{3.41}$$

$$\mathbf{B} \cdot \frac{\partial \mathbf{A}}{\partial t} = -\mathbf{B} \cdot \mathbf{E} + \mathbf{B} \cdot \nabla \psi = -\mathbf{B} \cdot \mathbf{E} + \nabla \cdot (\psi \mathbf{B}) , \tag{3.42}$$

where ψ is an electric potential. With Ohm's law, we finally obtain the local form of the magnetic helicity conservation:

$$\frac{\partial H^{\mathrm{m}}}{\partial t} = -\nabla \cdot \left(-(\mathbf{u} \times \mathbf{B}) \times \mathbf{A} + \frac{1}{ne} (\mathbf{j} \times \mathbf{B}) \times \mathbf{A} + \mu_0 \eta \mathbf{j} \times \mathbf{A} - \psi \mathbf{B} \right)$$
$$- 2\mu_0 \eta \, \mathbf{j} \cdot \mathbf{B} . \tag{3.43}$$

We have a single source term – of viscous[3] nature – whose sign is not defined. The integral form of the magnetic helicity conservation is written

$$\boxed{\frac{d}{dt} \iiint H^{\mathrm{m}} \, d\mathcal{V} = - \oiint \left[-(\mathbf{u} \times \mathbf{B}) \times \mathbf{A} + \frac{1}{ne} (\mathbf{j} \times \mathbf{B}) \times \mathbf{A} - \psi \mathbf{B} \right] \cdot \mathbf{n} \, d\mathcal{S} .}$$
$$\tag{3.44}$$

---

[3] We will also use sometimes the word viscous to designate the magnetic diffusion term.

If we consider the particular case of the volume $\mathcal{V}_\infty$, the two first fluxes vanish on the surface $\mathcal{S}_\infty$, but *a priori* not the last one unless we impose a constant electric potential on $\mathcal{S}_\infty$ (which is equivalent to saying that the vector potential is constant on $\mathcal{S}_\infty$). It is this assumption that is made in general in order to be able to say that the magnetic helicity is an invariant of the generalized ideal MHD.

3.6 Alfvén's Theorem

To establish Alfvén's theorem we shall proceed by steps. In all these steps we will assume that the conducting fluid is described by the standard ideal MHD.

3.6.1 Magnetic Flux Conservation

We shall demonstrate a remarkable property related to the topology of the magnetic field. To do this, let us consider a magnetic tube as shown in Figure 3.1, i.e. a structure bounded by the magnetic lines themselves so that **B** has no component perpendicular to the lateral boundary of the tube.

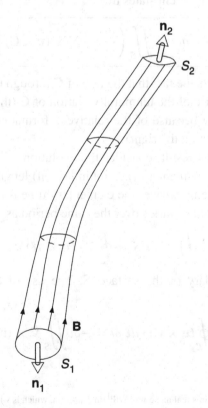

Figure 3.1 A magnetic tube.

We know that the magnetic flux crossing a closed surface (a tube with surfaces S_1 and S_2 in our case) is zero because $\nabla \cdot \mathbf{B} = 0$, and therefore

$$\Phi = \oiint \mathbf{B} \cdot \mathbf{n} \, d\mathcal{S} = \iint_{S_1} \mathbf{B} \cdot \mathbf{n}_1 \, d\mathcal{S} + \iint_{S_2} \mathbf{B} \cdot \mathbf{n}_2 \, d\mathcal{S} = 0 . \tag{3.45}$$

In our derivation, the unit vectors $\mathbf{n}_{1,2}$ perpendicular to the surfaces $S_{1,2}$ are oriented towards the exterior. If now we decide to take the direction of the magnetic field as a reference to compute the flux (i.e. $-\mathbf{n}_1$ is now taken; see Figure 3.1), we conclude immediately that the magnetic flux through any material surface S along a magnetic tube is conserved. Then, we may ask the following question: will the material volume initially located inside a magnetic tube remain stuck to the magnetic tube or will it move separately? To answer this question we need to use Kelvin's theorem.

3.6.2 Kelvin's Theorem

We shall establish Kelvin's theorem.[4] Suppose a material surface S moving with the fluid (i.e. a surface always composed of the same fluid particles). For any solenoidal field $\mathbf{C}$, this theorem states that

$$\frac{d}{dt} \iint_{S} \mathbf{C} \cdot d\mathcal{S} = \iint_{S} \left(\frac{\partial \mathbf{C}}{\partial t} - \nabla \times (\mathbf{u} \times \mathbf{C}) \right) \cdot d\mathcal{S} . \tag{3.46}$$

This relationship reflects the fact that the flux of $\mathbf{C}$ through the surface S can vary over time, firstly because of the temporal variation of $\mathbf{C}$ (the first contribution on the right) and secondly because of the relative deformation of S while moving (the second contribution on the right).

To demonstrate this result, consider the evolution of S in Figure 3.2. The surface adjacent to the elementary (i.e. infinitesimal) length $d\boldsymbol{\ell}$ increases during the elementary displacement (over the elementary time δt) of $\delta S = (\mathbf{u} \times d\boldsymbol{\ell})\delta t$. Thus, the flux of the field $\mathbf{C}$ varies over the same period as

$$\delta \iint_{S} \mathbf{C} \cdot d\mathcal{S} = \oint_{\mathcal{L}} \mathbf{C} \cdot (\mathbf{u} \times d\boldsymbol{\ell})\delta t , \tag{3.47}$$

where $\mathcal{L}$ is the boundary of the surface S. The use of Stokes' theorem (see Appendix 2) gives

$$\delta \iint_{S} \mathbf{C} \cdot d\mathcal{S} = -\oint_{\mathcal{L}} (\mathbf{u} \times \mathbf{C}) \cdot d\boldsymbol{\ell} \, \delta t = -\delta t \iint_{S} \nabla \times (\mathbf{u} \times \mathbf{C}) \cdot d\mathcal{S} , \tag{3.48}$$

---

[4] Kelvin's theorem (Lord Kelvin's real name was William Thomson, which is why sometimes one speaks also of Thomson's theorem) dates back to 1869 and concerns neutral fluids. In fact, Helmholtz had published in 1858 a similar theorem for vorticity lines (or tubes). Therefore, we sometimes refer to the Kelvin–Helmholtz theorem.

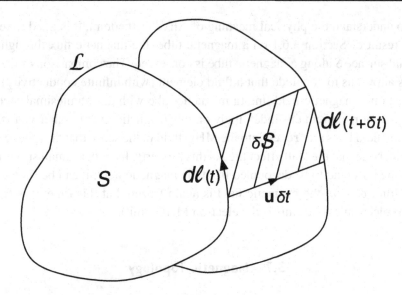

Figure 3.2 Movement of the material surface S in a time δt.

hence the evaluation of the flux variation of $\mathbf{C}$ due to the deformation of the surface:

$$\frac{\delta \iint_S \mathbf{C} \cdot d\mathcal{S}}{\delta t} \equiv \iint_S \mathbf{C} \cdot d\frac{\delta \mathcal{S}}{\delta t} = -\iint_S \nabla \times (\mathbf{u} \times \mathbf{C}) \cdot d\mathcal{S}, \qquad (3.49)$$

which demonstrates relation (3.46). To demonstrate Kelvin's theorem for the inviscid Navier–Stokes equations, we need only apply relation (3.46) to the field $\mathbf{C} = \mathbf{w}$. One obtains

$$\frac{d}{dt} \iint_S \mathbf{w} \cdot d\mathcal{S} = 0. \qquad (3.50)$$

Note that this theorem can also be written in terms of an integral over a closed curve (Kelvin's theorem).

3.6.3 Alfvén's Theorem

Now let us consider the case of the magnetic field. The application of the Kelvin–Helmholtz theorem to any material surface S moving with the MHD fluid gives trivially

$$\boxed{\frac{d}{dt} \iint_S \mathbf{B} \cdot d\mathcal{S} = 0.} \qquad (3.51)$$

Therefore, in the limit of a perfect conductor ($\eta \to 0$), the magnetic flux through any material surface advected by the fluid is conserved. This is **Alfvén's theorem**, which is also called the frozen-in theorem (Alfvén, 1942).

To understand the physical meaning of Alfvén's theorem, it is good to return to the result of Section 3.6.1 on a magnetic tube: the magnetic flux through any material surface S along a magnetic tube is conserved. The comparison of the two results allows us to conclude that a fluid element (with infinite conductivity) that is based on a magnetic field line (a magnetic tube with an infinitesimal section) at a given moment remains indefinitely on this magnetic line: in other words, the magnetic field lines are **frozen** into the MHD fluid in the sense that they move with the fluid. Note that the Hall effect breaks this property. Indeed, at small space-time scales the ions tend to be decoupled from the magnetic field: it can be shown that in the limit $\ell \ll d_i$ the magnetic field is again frozen, but this time, it is frozen into the electronic fluid (this is the electron MHD limit).

3.7 Magnetic Topology

Let us come back to the magnetic helicity and consider a standard ideal MHD fluid made only of a magnetic tube T. The conservation law of magnetic helicity can be written as

$$\frac{\partial H^{\mathrm{m}}}{\partial t} = \nabla \cdot [(\mathbf{u} \times \mathbf{B}) \times \mathbf{A} + \psi \mathbf{B}]$$
$$= \nabla \cdot [(\mathbf{A} \cdot \mathbf{u})\mathbf{B} - (\mathbf{A} \cdot \mathbf{B})\mathbf{u} + \psi \mathbf{B}] , \qquad (3.52)$$

hence we obtain the relationship

$$\frac{dH^{\mathrm{m}}}{dt} \equiv \frac{\partial H^{\mathrm{m}}}{\partial t} + \mathbf{u} \cdot \nabla H^{\mathrm{m}} = \nabla \cdot [(\mathbf{A} \cdot \mathbf{u})\mathbf{B} + \psi \mathbf{B}] - H^{\mathrm{m}} \nabla \cdot \mathbf{u}, \qquad (3.53)$$

where d/dt must be understood as the Lagrangian derivative. In the case of a closed magnetic tube T, the integral form reads

$$\iiint_{\mathrm{T}} \frac{dH^{\mathrm{m}}}{dt} \, d\mathcal{V} = \oiint [(\mathbf{A} \cdot \mathbf{u})\mathbf{B} + \psi \mathbf{B}] \cdot \mathbf{n} \, d\mathcal{S} - \iiint_{T} H^{\mathrm{m}} \nabla \cdot \mathbf{u} \, d\mathcal{V}$$
$$= - \iiint_{T} H^{\mathrm{m}} \nabla \cdot \mathbf{u} \, d\mathcal{V}. \qquad (3.54)$$

In the particular case of an incompressible fluid, we obtain the following law of conservation:

$$\boxed{\frac{d}{dt} \iiint_{\mathrm{T}} H^{\mathrm{m}} \, d\mathcal{V} = 0,} \qquad (3.55)$$

which means that the magnetic helicity in an incompressible closed magnetic tube is conserved over time.[5]

---

[5] The reader must pay attention to the non-trivial passage from a Lagrangian derivative inside the integral to outside. In this mathematical transformation we use the relation $d(d\mathcal{V})/dt = 0$ which holds for an incompressible fluid.

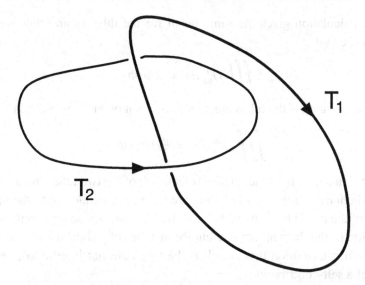

Figure 3.3 Magnetic helicity measures the degree of linkage of magnetic field lines.

Now let us consider the case of Figure 3.3, where two closed thin tubes (of infinitesimal section) are interlinked. We already know from Alfvén's theorem that the tubes will remain linked in the same manner for ever (or at least as long as the approximation of ideal standard MHD is applicable): one says that the interlinked magnetic tubes preserve their topology. Now we shall measure quantitatively this topology by using the magnetic helicity. The magnetic helicity of tube T_1 is

$$\iiint_{T_1} H_1^m \, d\mathcal{V} = \iiint_{T_1} \mathbf{A} \cdot \mathbf{B} \, d\ell \, d\mathcal{S} \tag{3.56}$$

$$= \iiint_{T_1} \mathbf{A} \cdot d\boldsymbol{\ell} \, B \, d\mathcal{S}$$

$$= \oint_{\mathcal{C}_1} \mathbf{A} \cdot \left(\iint_{\delta s_1} B \, d\mathcal{S} \right) d\boldsymbol{\ell} = \oint_{\mathcal{C}_1} \Phi_1 \mathbf{A} \cdot d\boldsymbol{\ell} = \Phi_1 \oint_{\mathcal{C}_1} \mathbf{A} \cdot d\boldsymbol{\ell},$$

where $\mathcal{C}_1$ is the contour defined by the tube T_1, δs_1 is the tube section, and Φ_1 is the magnetic flux associated with it; this flux is constant along the tube and therefore can be put outside of the integral. We still have to evaluate the integral of the vector potential over a closed contour. By noting that

$$\oint_{\mathcal{C}_1} \mathbf{A} \cdot d\boldsymbol{\ell} = \iint_{\mathcal{S}_1} \nabla \times \mathbf{A} \cdot \mathbf{n} \, d\mathcal{S} = \iint_{\mathcal{S}_1} \mathbf{B} \cdot \mathbf{n} \, d\mathcal{S} = \Phi_2, \tag{3.57}$$

where $\mathcal{S}_1$ is the surface defined by the tube, one finds

$$\iiint_{T_1} H_1^m \, d\mathcal{V} = \Phi_1 \Phi_2. \tag{3.58}$$

A similar calculation gives the same result for the tube T_2 and thus we finally obtain for the system

$$\iiint H_{tot}^{m} \, d\mathcal{V} = 2\Phi_1\Phi_2 \,. \tag{3.59}$$

It is easy to see that, if the tubes are wound on each other N times, then

$$\iiint H_{tot}^{m} \, d\mathcal{V} = \pm 2N\Phi_1\Phi_2 \,, \tag{3.60}$$

where the sign depends on the relative orientation of the two tubes. As a result, the magnetic helicity and the magnetic flux are intimately connected to the magnetic topology of the MHD fluid. It can be seen that our calculation gives only a mutual contribution of the flux: in fact, the choice of tubes of infinitesimal section leads trivially to a contribution that is null if the tubes are not interlinked, hence the absence of a self-contribution.

To conclude, we give an example of magnetic topology at MHD scales observed on the Sun: Figure 3.4 displays a network of coronal loops in an active

Figure 3.4 Coronal loops observed with NASA's space telescope TRACE at the wavelength 17.1 nm (June 6, 1999). The study of the magnetic topology on the Sun is important to understand, for example, the magnetic reconnection processes. Credit: NASA/TRACE.

region. These loops are at a temperature greater than a million degrees. In this medium, the approximation of ideal MHD is believed to break only at scales smaller than 1 km. At such small scales the magnetic topology can be modified via, for example, a magnetic reconnection (see Chapter 7). More information about the topological properties of magnetic helicity can be found e.g. in Berger and Field (1984) or in Moffatt (1978).

3.8 Topology at Sub-ion Scales

In the context of standard incompressible MHD we have seen that the magnetic helicity provides an interesting tool to quantity the topology (i.e. the self-linkage or knottedness) of a magnetic tube. However, at sub-ion scales ($\ell < d_i$) the Hall effect breaks the properties described above because basically the ions and the electrons start to be decoupled, and only electrons remain frozen into the magnetic field. We can prove, however, that the topology is preserved for a generalized vorticity tube by using the generalized helicity. By definition the generalized vorticity is

$$\mathbf{\Omega} = \mathbf{b} + d_i \mathbf{w}, \tag{3.61}$$

where the magnetic field is normalized with respect to a velocity (this is the incompressible Hall MHD approximation). It is related to the generalized vector potential Υ through the relation

$$\mathbf{\Omega} = \nabla \times \Upsilon, \tag{3.62}$$

with $\Upsilon = \mathbf{a} + d_i \mathbf{u}$ and $\mathbf{b} = \nabla \times \mathbf{a}$. The generalized helicity is then defined as

$$H^g = \Upsilon \cdot \mathbf{\Omega}. \tag{3.63}$$

By analogy with the magnetic field in standard MHD we can easily show that the flux is conserved along a generalized vorticity tube (see Figure 3.5) since $\nabla \cdot \mathbf{\Omega} = 0$. Then, after a few calculations we see that

$$\frac{\partial \mathbf{\Omega}}{\partial t} = \nabla \times (\mathbf{u} \times \mathbf{\Omega}), \tag{3.64}$$

and thus thanks to Kelvin's theorem we can conclude (Turner, 1986) that

$$\boxed{\frac{d}{dt} \iint_S \mathbf{\Omega} \cdot d\mathcal{S} = 0.} \tag{3.65}$$

Therefore, in the limit of an inviscid and ideal fluid ($\nu \to 0$ and $\eta \to 0$), the generalized vorticity flux through any material surface advected by the fluid is conserved. Likewise in standard MHD, we can conclude that a fluid element based on a generalized vorticity field line (a tube with an infinitesimal section) at a given

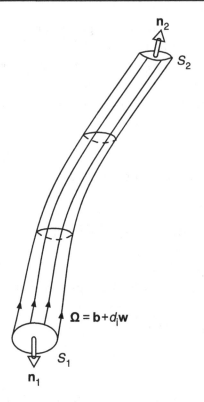

Figure 3.5 A generalized vorticity tube.

moment remains indefinitely on this line. In other words, the generalized vorticity field lines are frozen into the Hall MHD fluid in the sense that they move with the fluid. This is the generalization of Alfvén's theorem to Hall MHD.

We can evaluate the generalized helicity inside a generalized vorticity tube T. A simple calculation gives

$$\frac{d}{dt} \iiint_{\mathrm{T}} H^{\mathrm{g}}\, d\mathcal{V} = 0, \tag{3.66}$$

which means that in an incompressible closed generalized vorticity tube H^{g} is conserved over time. Finally, let us consider the case of Figure 3.3 for two closed thin generalized vorticity tubes (of infinitesimal section). A classical calculation gives

$$\iiint H^{\mathrm{g}}_{\mathrm{tot}}\, d\mathcal{V} = 2\Phi_1 \Phi_2, \tag{3.67}$$

where Φ_i must be understood as the flux of the generalized vorticity in a section of tube i. Therefore, we see that the generalized helicity and the generalized vorticity flux are intimately connected to the topology of the Hall MHD fluid. This topology is preserved in an inviscid and ideal fluid.

Exercises for Part I

I.a Kinetic Helicity

What is the local form of the equation of kinetic helicity conservation,

$$H^u \equiv \mathbf{u} \cdot (\nabla \times \mathbf{u}),$$

in incompressible MHD? In what limit does the kinetic helicity become an inviscid invariant?

I.b Cross-helicity and Topology

Calculate the cross-helicity of a magnetic tube T_1 interlinked with a vorticity tube T_2 (see Figure I.b.1). What is your conclusion?

I.c Reduced MHD Approximation

Consider a compressible MHD flow subjected to a strong magnetic field $\mathbf{B}_0$. We introduce the following dimensionless quantities:

$$\mathbf{B} \rightarrow B_0 \mathbf{e}_z + \epsilon B_0 \mathbf{B}, \quad \mathbf{u} \rightarrow \epsilon b_0 \mathbf{u}, \quad \rho \rightarrow \rho_0 \rho,$$

where ϵ is a small parameter ($\epsilon \ll 1$), ρ_0 is a constant density, and b_0 is the Alfvén speed ($b_0 \equiv B_0/\sqrt{\mu_0 \rho_0}$). Demonstrate, by a perturbative development at the main order, that the MHD equations reduce to the following system involving only the transverse field components:

$$\frac{\partial \mathbf{u}_\perp}{\partial t} + \mathbf{u}_\perp \cdot \nabla_\perp \mathbf{u}_\perp - \frac{\partial \mathbf{b}_\perp}{\partial z} = -\nabla_\perp \left(\frac{b_\perp^2}{2} \right) + \mathbf{b}_\perp \cdot \nabla_\perp \mathbf{b}_\perp + \nu \Delta \mathbf{u}_\perp,$$

$$\frac{\partial \mathbf{b}_\perp}{\partial t} + \mathbf{u}_\perp \cdot \nabla_\perp \mathbf{b}_\perp - \frac{\partial \mathbf{u}_\perp}{\partial z} = \mathbf{b}_\perp \cdot \nabla_\perp \mathbf{u}_\perp + \eta \Delta \mathbf{b}_\perp.$$

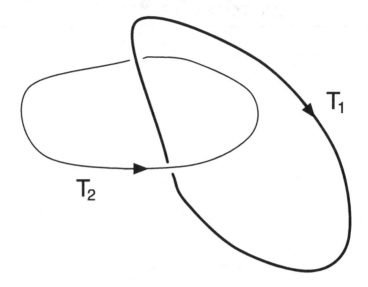

Figure I.b.1 We assume that T_1 is a magnetic tube and T_2 a vorticity tube.

This system corresponds to the reduced MHD equations. Although it was initially developed in the context of tokamaks (Strauss, 1976; Zank and Matthaeus, 1992), the reduced MHD is now used in astrophysics, e.g. for the modeling of magnetic loops in the solar corona (Buchlin and Velli, 2007).

I.d Generalized Helicity

There is a third (and last) inviscid and ideal invariant in Hall MHD; it is the generalized helicity (Turner, 1986),

$$H^g = (\mathbf{a} + d_i \mathbf{u}) \cdot (\mathbf{b} + d_i \nabla \times \mathbf{u}) ,$$

where $\mathbf{b} = \nabla \times \mathbf{a}$. Prove this result in the incompressible case.

Part II

Fundamental Processes

4

Magnetohydrodynamic Waves

In contrast to a mean flow in hydrodynamics, which can be eliminated by a Galilean transformation, in MHD the presence of a uniform magnetic field has a deep influence on the linear and nonlinear dynamics. One consequence is the presence of waves in incompressible MHD. These are called Alfvén waves. This chapter shall describe the different types of waves encountered in generalized MHD, starting with the simplest case: the incompressible standard MHD.

4.1   Magnetic Tension

To understand the origin of Alfvén waves, it is relevant to define the magnetic tension. As we know, the Laplace force is

$$\mathbf{F} = \mathbf{j} \times \mathbf{B} = \frac{1}{\mu_0}(\nabla \times \mathbf{B}) \times \mathbf{B}. \tag{4.1}$$

With the vectorial identity

$$\nabla(\mathbf{a} \cdot \mathbf{b}) = (\mathbf{a} \cdot \nabla)\mathbf{b} + (\mathbf{b} \cdot \nabla)\mathbf{a} + \mathbf{a} \times (\nabla \times \mathbf{b}) + \mathbf{b} \times (\nabla \times \mathbf{a}) \tag{4.2}$$

we obtain

$$\nabla(B^2) = 2(\mathbf{B} \cdot \nabla)\mathbf{B} + 2\mathbf{B} \times (\nabla \times \mathbf{B}), \tag{4.3}$$

and thus

$$\mathbf{F} = \frac{1}{\mu_0}(\nabla \times \mathbf{B}) \times \mathbf{B} = -\frac{1}{2}\nabla\left(\frac{B^2}{\mu_0}\right) + (\mathbf{B} \cdot \nabla)\frac{\mathbf{B}}{\mu_0}. \tag{4.4}$$

Then, the equation of motion can be written as

$$\rho\left(\frac{\partial \mathbf{u}}{\partial t} + \mathbf{u} \cdot \nabla \mathbf{u}\right) = -\nabla\left(P + \frac{B^2}{2\mu_0}\right) + (\mathbf{B} \cdot \nabla)\frac{\mathbf{B}}{\mu_0}. \tag{4.5}$$

51

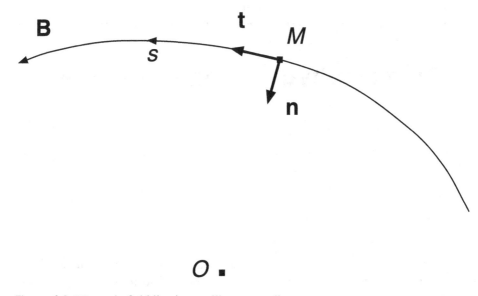

Figure 4.1 Magnetic field line in curvilinear coordinates.

The pressure is expressed as the sum of two pressures: the standard pressure plus a **magnetic pressure**. The second term on the right-hand side is a **magnetic tension.** To better understand the role of each term, it is relevant to use the local coordinates of Figure 4.1, where **t** is the tangent, **n** the normal and s the curvilinear coordinate; $R = OM$ is the curvature radius. We obtain the following expression:

$$(\mathbf{B} \cdot \nabla)\mathbf{B} = B\frac{\partial B\mathbf{t}}{\partial s} = \frac{\partial B^2/2}{\partial s}\mathbf{t} + B^2\frac{\partial \mathbf{t}}{\partial s}$$

$$= \frac{\partial B^2/2}{\partial s}\mathbf{t} + \frac{B^2}{R}\frac{\partial \mathbf{t}}{\partial \theta} = \frac{\partial B^2/2}{\partial s}\mathbf{t} + \frac{B^2}{R}\mathbf{n}, \qquad (4.6)$$

where

$$\mathbf{F} = \frac{1}{2\mu_0}\frac{\partial B^2}{\partial s}\mathbf{t} + \frac{B^2}{R\mu_0}\mathbf{n} - \nabla\left(\frac{B^2}{2\mu_0}\right). \qquad (4.7)$$

Therefore, the Laplace force involves a magnetic pressure and two magnetic tensions: a tension along **B** and another perpendicular to **B** which tends to reduce the curvature. Figure 4.2 gives an illustration of the immediate consequence of this mechanism: the generation of an Alfvén wave. Indeed, the passage of a perturbation leads to the curvature of the magnetic field lines because it is frozen into the fluid (Alfvén's theorem), and then a normal magnetic tension appears. As the curvature increases the restoring magnetic force rises and eventually the inertia of the fluid is overcome. Then, the flow goes back to the left carrying the magnetic field lines and the process continues until the dissipation entirely damps these oscillations. Finally, incompressible waves can propagate along these magnetic lines (or magnetic threads): these are Alfvén waves that propagate at the

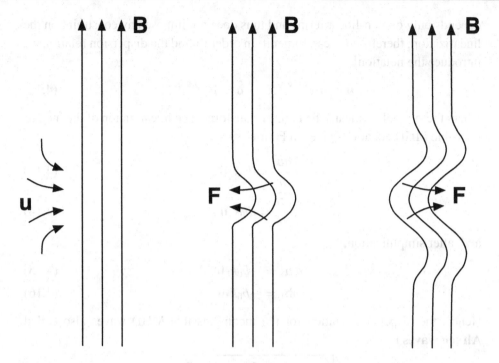

Figure 4.2 Magnetic field evolution after the passage of a perturbation. The oscillations caused by the disturbance give birth to incompressible MHD waves: the Alfvén waves.

Alfvén speed (see Section 4.2). In the next section, a rigorous derivation of the Alfvén waves is given.

4.2 Alfvén Waves

Let us consider the standard incompressible, ideal, and inviscid MHD. We introduce a uniform magnetic field $\mathbf{B}_0 = B_0 \mathbf{e}_\parallel$ ($\mathbf{e}_\parallel$ is a unit vector). We also introduce the reduced form of the magnetic field (normalized with respect to a velocity),

$$\mathbf{b}_0 \equiv \frac{\mathbf{B}_0}{\sqrt{\mu_0 \rho_0}}, \tag{4.8}$$

and obtain

$$\frac{\partial \mathbf{u}}{\partial t} + \mathbf{u} \cdot \nabla \mathbf{u} = -\nabla P_* + (\mathbf{b} + \mathbf{b}_0) \cdot \nabla \mathbf{b}, \tag{4.9}$$

$$\frac{\partial \mathbf{b}}{\partial t} + \mathbf{u} \cdot \nabla \mathbf{b} = (\mathbf{b} + \mathbf{b}_0) \cdot \nabla \mathbf{u}, \tag{4.10}$$

$$\nabla \cdot \mathbf{u} = \nabla \cdot \mathbf{b} = 0, \tag{4.11}$$

where the total pressure can be written as $P_* = P/\rho_0 + (\mathbf{b} + \mathbf{b}_0)^2/2$. If we apply the divergence operator on Eq. (4.9), we see immediately that the total pressure depends only on non-linear terms and thus does not impose any constraint on the linearization; therefore we can forget it. In order to find the dispersion relation we introduce the notation[1]

$$\hat{\mathbf{u}} = \mathbf{u}_1 e^{i(\mathbf{k} \cdot \mathbf{x} - \omega t)}, \quad \hat{\mathbf{b}} = \mathbf{b}_1 e^{i(\mathbf{k} \cdot \mathbf{x} - \omega t)}, \tag{4.12}$$

where the symbol $\hat{}$ means the Fourier transform. The linearization of the incompressible MHD equations gives in Fourier space

$$\frac{\partial \hat{\mathbf{u}}}{\partial t} = i b_0 k_\parallel \hat{\mathbf{b}}, \tag{4.13}$$

$$\frac{\partial \hat{\mathbf{b}}}{\partial t} = i b_0 k_\parallel \hat{\mathbf{u}}, \tag{4.14}$$

and, after simplification,

$$\omega \mathbf{u}_1 = -b_0 k_\parallel \mathbf{b}_1, \tag{4.15}$$

$$\omega \mathbf{b}_1 = -b_0 k_\parallel \mathbf{u}_1. \tag{4.16}$$

Hence, the dispersion relation for the incompressible MHD waves also called **Alfvén waves** is

$$\boxed{\omega^2 = (b_0 k_\parallel)^2 = (\mathbf{b}_0 \cdot \mathbf{k})^2.} \tag{4.17}$$

There are transverse waves ($\hat{\mathbf{u}}$ and $\hat{\mathbf{b}} \perp \mathbf{k}$) which propagate at the group speed b_0 – the Alfvén speed – with the phase speed $u_\phi = b_0 k_\parallel / k$ (semi-dispersive waves). These are basically anisotropic waves. In the solar corona, the Alfvén speed can reach 10^4 km/s (with $B_0 \sim 10^{-2}$ T and $\rho_0 \sim 10^{-12}$ kg/m^3) and in the solar wind (at 1 AU) it is approximately 10 km/s (with $B_0 \sim 10^{-9}$ T and $\rho_0 \sim 10^{-20}$ kg/m^3). It is thought that these waves play a major role in the solar coronal heating (Heyvaerts and Priest, 1983; Bigot *et al.*, 2008a; Chandran, 2010). Note that a distinction is sometimes made between the fluctuations transverse and parallel to $\mathbf{b}_0$: the former are called **shear-Alfvén** waves and the latter **pseudo-Alfvén** waves.

The theoretical prediction[2] of Alfvén waves dates back to 1942 (Alfvén, 1942; Falthammar, 2007), but one had to wait until 1959 for the (first clear) experimental verification (Allen *et al.*, 1959): as shown in Figure 4.3 the speed of propagation of the Alfvén waves produced in the experimental setup (an ionized hydrogen tube) follows quite well a law proportional to the applied magnetic field. Note

---

[1] Note that we can also introduce the D'Alembert operator to solve the problem. In this chapter we will follow the classical technique and use the Fourier transform.

[2] Despite the relative simplicity of the demonstration made in 1942, H. Alfvén had a lot of difficulties in convincing member of the scientific community, who thought that J. C. Maxwell would have found this result if it was relevant. The situation changed after a seminar given in Chicago and in the presence of E. Fermi (Nobel Prize in Physics 1938) when he said "Of course, such waves could exist!". Then, the community said "Oh. Of course ..."

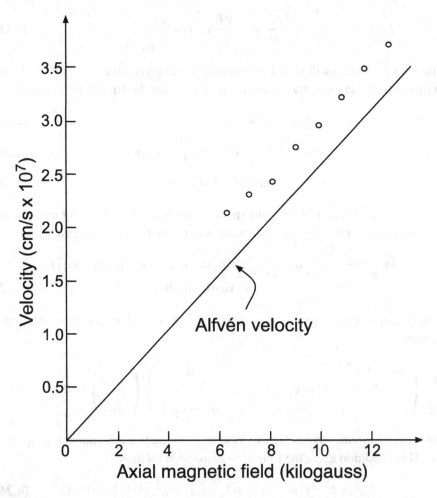

Figure 4.3 Experimental evidence of Alfvén waves in a tube of length 34 inches and diameter $5\frac{3}{4}$ inches filled with ionized hydrogen: the group speed measured (circles) is – as expected – proportional to the applied magnetic field (the systematic bias is due to instruments). Adapted from Allen *et al.* (1959).

that the first experiment devoted to the detection of Alfvén waves was performed in 1949 with liquid mercury (Lundquist, 1949). The subsequent experiments done in 1952 and 1954, with ionized helium and liquid sodium, respectively, were more conclusive (Bostick and Levine, 1952; Lehnert, 1954). For the Sun, it was only in 2008 that the first signatures of Alfvén waves were obtained with the data from the space telescope Hinode/JAXA (De Pontieu *et al.*, 2007).

4.3 Magnetosonic Waves

Let us consider now the general case of compressible (still standard) MHD. We use a polytropic closure for which the pressure equation is

$$\frac{\partial P}{\partial t} = -\frac{\gamma P}{\rho} \nabla \cdot (\rho \mathbf{u}) \,. \tag{4.18}$$

Following the same method as in the previous section (with $\hat{P} = P_1 e^{i(\mathbf{k}\cdot\mathbf{x}-\omega t)}$), we obtain the following linearized system (ρ_0 and P_0 are the equilibrium values):

$$\omega P_1 = \rho_0 c_S^2 \, \mathbf{k} \cdot \mathbf{u}_1 \,, \tag{4.19}$$

$$\omega \mathbf{u}_1 = \mathbf{k} \left(\frac{P_1}{\rho_0} + \mathbf{b}_0 \cdot \mathbf{b}_1 \right) - b_0 k_\parallel \mathbf{b}_1 \,, \tag{4.20}$$

$$\omega \mathbf{b}_1 = -b_0 k_\parallel \mathbf{u}_1 + \mathbf{b}_0 (\mathbf{k} \cdot \mathbf{u}_1) \,, \tag{4.21}$$

where $c_S \equiv \sqrt{\gamma P_0 / \rho_0}$ is the **sound speed**. We rewrite Eq. (4.20) by introducing the expressions for the perturbed pressure and magnetic field. We easily find

$$\left[\omega^2 - (\mathbf{k} \cdot \mathbf{b}_0)^2 \right] \mathbf{u}_1 = \left[(c_S^2 + b_0^2)(\mathbf{k} \cdot \mathbf{u}_1) - (\mathbf{u}_1 \cdot \mathbf{b}_0)(\mathbf{k} \cdot \mathbf{b}_0) \right] \mathbf{k}$$
$$- (\mathbf{k} \cdot \mathbf{b}_0)(\mathbf{k} \cdot \mathbf{u}_1) \mathbf{b}_0 \,. \tag{4.22}$$

Without loss of generality we may define $\mathbf{k} = k_\perp \mathbf{e}_y + k_\parallel \mathbf{e}_z$ and thus obtain the expression

$$\begin{pmatrix} \omega^2 - k_\parallel^2 b_0^2 & 0 & 0 \\ 0 & \omega^2 - k_\perp^2 c_S^2 - k^2 b_0^2 & -k_\perp k_\parallel c_S^2 \\ 0 & -k_\perp k_\parallel c_S^2 & \omega^2 - k_\parallel^2 c_S^2 \end{pmatrix} \begin{pmatrix} u_{1x} \\ u_{1y} \\ u_{1z} \end{pmatrix} = 0 \,. \tag{4.23}$$

A non-trivial solution exists for this system if the matrix determinant is equal to zero. This condition gives the following dispersion relation:

$$(\omega^2 - k_\parallel^2 b_0^2) \left[\omega^4 - (c_S^2 + b_0^2) k^2 \omega^2 + k^2 c_S^2 k_\parallel^2 b_0^2 \right] = 0 \,. \tag{4.24}$$

An immediate solution of this system is the Alfvén wave that we have already derived in the incompressible case; thus this wave is not affected by compressibility. In addition, we have two new solutions, namely the fast and slow **magnetosonic waves** for which, respectively,

$$\omega_+^2 = \frac{k^2}{2} \left[c_S^2 + b_0^2 + \sqrt{(c_S^2 + b_0^2)^2 - 4 c_S^2 b_0^2 k_\parallel^2 / k^2} \right] \,, \tag{4.25}$$

$$\omega_-^2 = \frac{k^2}{2} \left[c_S^2 + b_0^2 - \sqrt{(c_S^2 + b_0^2)^2 - 4 c_S^2 b_0^2 k_\parallel^2 / k^2} \right] \,. \tag{4.26}$$

It is interesting to look at two particular limits, $k = k_\parallel$ and $k = k_\perp$. In the first case, one of the two waves can be identified as the classical – thus longitudinal (see relation (4.22)) – sound wave, whereas the other wave can be identified as the classical – thus transverse – Alfvén wave. In the second case, only the fast

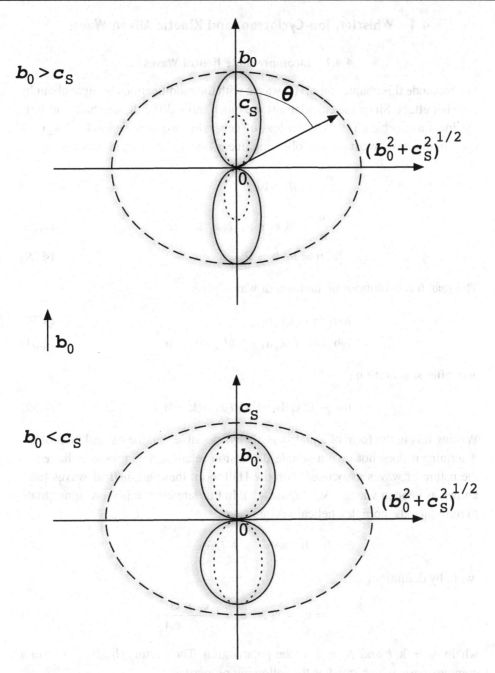

Figure 4.4 Phase velocity representation in polar coordinates of the three MHD waves: Alfvén (solid), fast (dash), and slow (dot) magnetosonic waves. By definition θ is the angle between $\mathbf{b}_0$ and $\mathbf{k}$.

wave survives: it is a longitudinal wave which looks like a modified sound wave. These results are summarized in Figure 4.4, in which the phase velocities of the three MHD waves are plotted for a medium where $b_0 > c_S$ (top) and for $b_0 < c_S$ (bottom).

4.4   Whistler, Ion-Cyclotron, and Kinetic Alfvén Waves

4.4.1   Incompressible Helical Waves

We conclude this chapter on MHD waves with the modifications brought about by the Hall effect. Since the calculations are much more difficult, we shall start with the incompressible case. Without loss of generality we may define $\mathbf{k} = k_\perp \mathbf{e}_y + k_\parallel \mathbf{e}_z$. In this case, we have the following linear (ideal and inviscid) system:

$$\frac{\partial \hat{\mathbf{u}}}{\partial t} = i b_0 k_\parallel \hat{\mathbf{b}}, \tag{4.27}$$

$$\frac{\partial \hat{\mathbf{b}}}{\partial t} = i b_0 k_\parallel \hat{\mathbf{u}} + d_i b_0 k_\parallel (\mathbf{k} \times \hat{\mathbf{b}}), \tag{4.28}$$

$$\mathbf{k} \cdot \hat{\mathbf{u}} = \mathbf{k} \cdot \hat{\mathbf{b}} = 0. \tag{4.29}$$

The search for solutions in the form of waves gives

$$\omega \mathbf{u}_1 = -b_0 k_\parallel \mathbf{b}_1, \tag{4.30}$$

$$\omega \mathbf{b}_1 = -b_0 k_\parallel \mathbf{u}_1 + i d_i b_0 k_\parallel (\mathbf{k} \times \mathbf{b}_1), \tag{4.31}$$

and, after substitution,

$$(\omega^2 - b_0^2 k_\parallel^2) \mathbf{b}_1 = i d_i b_0 k_\parallel \omega (\mathbf{k} \times \mathbf{b}_1). \tag{4.32}$$

Writing this in the form of a matrix is always possible, but the cancellation of the determinant does not give a simple dispersion relation. The reason is linked to the nature of waves associated with the Hall term: these are **helical waves** (also known as helicon waves – see Figure 4.5). In this situation, it is more appropriate to introduce the complex helical vectors

$$\mathbf{h}_\Lambda(\mathbf{k}) = \mathbf{e}_\theta + i \Lambda \mathbf{e}_\phi, \tag{4.33}$$

with, by definition,

$$\mathbf{e}_\theta = \mathbf{e}_\phi \times \mathbf{e}_k, \qquad \mathbf{e}_\phi = \frac{\mathbf{e}_\parallel \times \mathbf{e}_k}{|\mathbf{e}_\parallel \times \mathbf{e}_k|}, \tag{4.34}$$

where $\mathbf{e}_k = \mathbf{k}/k$ and $\Lambda = \pm$ is the polarization. The vectors $(\mathbf{h}_+, \mathbf{h}_-, \mathbf{e}_k)$ form a complex basis which satisfies the following properties:

$$\mathbf{h}_{-\Lambda}(-\mathbf{k}) = \mathbf{h}_\Lambda(\mathbf{k}), \qquad \mathbf{e}_k \times \mathbf{h}_\Lambda = -i \Lambda \, \mathbf{h}_\Lambda, \tag{4.35}$$

$$\mathbf{k} \cdot \mathbf{h}_\Lambda = 0, \qquad \mathbf{h}_\Lambda \cdot \mathbf{h}_{\Lambda'} = 2 \delta_{-\Lambda' \Lambda}. \tag{4.36}$$

We define

$$\mathbf{b}_1 = \mathcal{B}_+ \mathbf{h}_+ + \mathcal{B}_- \mathbf{h}_-, \tag{4.37}$$

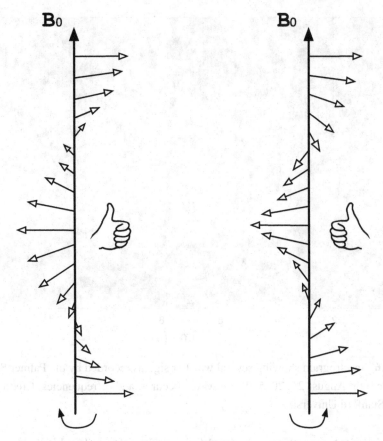

Figure 4.5 Propagation of helical waves with right (right) and left (left) circular polarities. In order to find the direction of rotation one can use the right- (or left-)hand rule.

and we get after projection

$$(\omega^2 - b_0^2 k_\parallel^2)(\mathcal{B}_+ \mathbf{h}_+ + \mathcal{B}_- \mathbf{h}_-) = d_i b_0 k_\parallel k \omega (\mathcal{B}_+ \mathbf{h}_+ - \mathcal{B}_- \mathbf{h}_-). \tag{4.38}$$

Relation (4.38) gives trivially (because we have diagonalized the system) the dispersion relation

$$\omega^2 - \Lambda d_i b_0 k_\parallel k \omega - b_0^2 k_\parallel^2 = 0, \tag{4.39}$$

for which the solutions are

$$\omega_\Lambda^s = \frac{s k_\parallel k d_i b_0}{2} \left(s\Lambda + \sqrt{1 + \frac{4}{d_i^2 k^2}} \right), \tag{4.40}$$

with s the directional polarity (by definition $s k_\parallel \geq 0$) and ω_Λ^s a positive defined quantity. With these notations, Λ gives an information about the wave

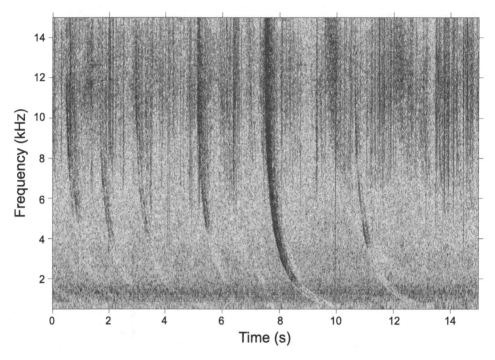

Figure 4.6 Spectrogram showing several whistler signals recorded by the Palmer Station (Antarctica) on August 24, 2005. These waves occur at audio frequencies. Credit: D. I. Golden, Stanford University.

polarization: if $\Lambda = s$, the wave is **right** circularly polarized and if $\Lambda = -s$ it is **left** circularly polarized.

In this incompressible limit both waves are transverse (to **k**); they are also dispersive. It is relevant to study the large ($kd_i \rightarrow 0$) and small ($kd_i \rightarrow +\infty$) scale limits. In the first case we obtain – as expected – the standard Alfvén wave: there is degeneracy. In the second case we obtain the whistler[3] wave, which is also called the incompressible **Alfvén–whistler** wave, with

$$\omega_s^{\mathrm{s}} = sk_{\parallel}kd_i b_0 \,, \tag{4.41}$$

and the **ion-cyclotron** wave with

$$\omega_{-s}^{\mathrm{s}} = \frac{sk_{\parallel}b_0}{d_i k} = \alpha\omega_{\mathrm{ci}} \,, \tag{4.42}$$

where $\alpha \equiv sk_{\parallel}/k = \cos\theta$ (θ is the angle between $\mathbf{b}_0$ and $\mathbf{k}$) and $\omega_{\mathrm{ci}} \equiv b_0/d_i$ is the ion-cyclotron (angular) frequency. For the ion-cyclotron wave there exists

---

[3] Whistler waves were first detected during World War 1. They are audio frequency waves often produced by lightning. Once produced in the magnetosphere these waves travel along closed magnetic field lines from one hemisphere to the other. Their phase and group velocities are both proportional to k, implying that higher-frequency waves have higher group and phase velocities. Thus, the high-frequency part of the whistler wave-packet will reach a detector earlier than its low-frequency part, and it will appear as a falling tone in a frequency–time sonogram (see Figure 4.6).

a **resonance** (in the asymptotic sense) which depends on the orientation of the wavevector: in the particular case where $k = k_\parallel$, the resonance appears exactly at the frequency ω_{ci}. In this fluid approach the resonance does not imply exchange of energy with particles and therefore there is no physical reason to describe these waves in the large-kd_i limit since then kinetic effects have to be taken into account. This constitutes the first limitation of the Hall MHD model. At higher frequencies only incompressible Alfvén–whistler waves survive, but they do not encounter any resonance: therefore, the Hall MHD approach does not become physically relevant at such high frequencies.[4] This is the second limitation of the model. The two dispersive branches are plotted on Figure 4.7 in normalized variables. Despite these two limits, Hall MHD is an interesting model because it allows us to include in a simple way the first decoupling effects between ions and electrons without entering into the difficult formulation of kinetic models.

4.4.2   Compressible Hall MHD Waves

We conclude this section with the most general case of compressible Hall MHD. In a classical manner, we obtain

$$\omega P_1 = \rho_0 c_S^2 \, \mathbf{k} \cdot \mathbf{u}_1, \tag{4.43}$$

$$\omega \mathbf{u}_1 = \mathbf{k} \left(\frac{P_1}{\rho_0} + \mathbf{b}_0 \cdot \mathbf{b}_1 \right) - b_0 k_\parallel \mathbf{b}_1, \tag{4.44}$$

$$\omega \mathbf{b}_1 = -b_0 k_\parallel \mathbf{u}_1 + \mathbf{b}_0 (\mathbf{k} \cdot \mathbf{u}_1) + i d_i b_0 k_\parallel (\mathbf{k} \times \mathbf{b}_1). \tag{4.45}$$

The application of the divergence operator on Eq. (4.44) gives the following condition:

$$(\omega^2 - k^2 c_S^2)(\mathbf{k} \cdot \mathbf{u}_1) = \omega k^2 (\mathbf{b}_1 \cdot \mathbf{b}_0), \tag{4.46}$$

which has to be used in order to reduce the system. After some manipulations (pressure substitution and use of the previous condition), we obtain the vectorial equation

$$(\Omega^2 - 1) \left(\Omega^2 - \frac{\beta}{\alpha^2} \right) \mathbf{b}_1 = -\Omega^2 \frac{b_{1\parallel}}{\alpha} \mathbf{e_k} + \frac{\Omega^2}{\alpha^2} b_{1\parallel} \mathbf{e_\parallel}$$

$$+ i d_i \Omega \left(\Omega^2 - \frac{\beta}{\alpha^2} \right) (\mathbf{k} \times \mathbf{b}_1), \tag{4.47}$$

where by definition $\Omega \equiv \omega/(k_\parallel b_0)$ is the normalized frequency, $\beta \equiv c_S^2/b_0^2$, and $\alpha \equiv \cos\theta$ (with θ the angle between $\mathbf{b}_0$ and $\mathbf{k}$). The reader is invited to write the matrix form for this equation for which the cancellation condition of the determinant gives finally the dispersion relation

---

[4] The fluid approach may be refined if we consider in the generalized form of Ohm's law an additional term which takes into account the electron inertia. The new system is then able to produce a resonance (i.e. an asymptote) around the electron-cyclotron (angular) frequency.

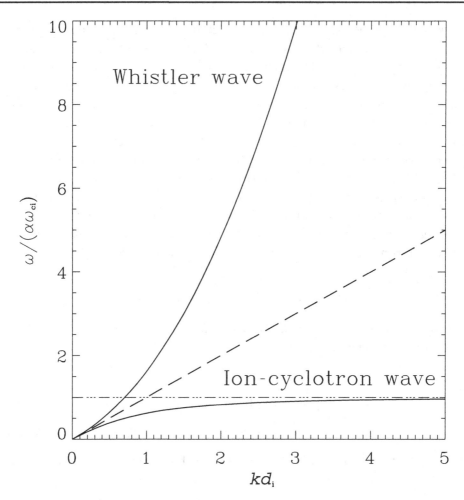

Figure 4.7 Dispersion relation (4.40) of incompressible Hall MHD in normalized variables (see the text): the left polarized wave tends to the ion-cyclotron wave when approaching the asymptote $\omega = \alpha\,\omega_{ci}$ (dot–dash line) whereas the right polarized wave tends to the whistler wave. The long-dash line corresponds to the dispersion relation of the Alfvén wave.

$$\Omega^6 - \left(1 + \frac{\beta+1}{\alpha^2} + K^2\right)\Omega^4 + \left(\frac{1 + 2\beta + \beta K^2}{\alpha^2}\right)\Omega^2 - \frac{\beta}{\alpha^2} = 0, \quad (4.48)$$

where $K \equiv kd_i$ is the normalized wavenumber. The first limit that we shall investigate is $\beta \gg 1$, i.e. the incompressible limit (compressible fluctuations propagate instantaneously). We easily show that the simplified system can be written exactly as relation (4.39). Thus, this limit corresponds to the waves discussed above in Section 4.4.1. The second interesting limit is $K \ll 1$, i.e. the standard MHD limit. Here again we can show that the dispersion relation may be reduced to the well-known form (4.24): we recover the magnetosonic waves. The third particular limit is the one for which we have a strictly parallel propagation ($\alpha = 1$). We see that $\Omega^2 = \beta$ is the solution: this is a sonic wave ($\omega^2 = k_\parallel^2 c_S^2$) which can be identified as a fast (if $\beta > 1$) or a slow (if $\beta < 1$) wave. Then, the dispersion relation reduces to

$$\left(\Omega^2 - \beta\right)\left(\Omega^4 - \left(2 + K^2\alpha^2\right)\Omega^2 + 1\right) = 0. \tag{4.49}$$

We can easily show that the cancellation of the fourth-order polynomial may be rewritten exactly as relation (4.39) for parallel propagation: we recover the incompressible modes with right or left polarization as particular solutions of the compressible case in parallel propagation. We can investigate a last relevant limit by writing the dispersion relation (4.48) in terms of the normalized frequency $\tilde{\Omega} \equiv \omega/\omega_{\mathrm{ci}} = \Omega K\alpha$; we obtain

$$\tilde{\Omega}^6 - \left(1 + \frac{\beta + 1}{\alpha^2} + K^2\right)K^2\alpha^2\tilde{\Omega}^4 + \left(1 + 2\beta + \beta K^2\right)K^4\alpha^2\tilde{\Omega}^2 - \beta K^6\alpha^4 = 0. \tag{4.50}$$

In the limits $\alpha \ll 1$ and $\tilde{\Omega} \ll 1$, the expression reduces to

$$(\beta + 1)\tilde{\Omega}^4 - \left(1 + 2\beta + \beta K^2\right)K^2\alpha^2\tilde{\Omega}^2 + \beta K^4\alpha^4 = 0, \tag{4.51}$$

whose solutions for $K > 1$ are approximatively $\tilde{\Omega} \simeq \sqrt{\beta/(1 + \beta)}K^2\alpha$ and $\tilde{\Omega} \simeq \alpha$. The first solution is the most interesting – this is the **kinetic Alfvén wave** with a right polarization[5] – whereas the second solution is nothing other than the ion-cyclotron Alfvén wave. For the general case, the study of compressible Hall MHD modes requires one to solve numerically the dispersion relation (see Figure 4.8).

Recent theoretical developments have led to a refined formulation of compressible Hall MHD in which some kinetic effects have been incorporated by modeling the linear Landau damping: this is called the Landau fluid model. Basically, in this approach a closure of the kinetic equations is used at the level of the heat fluxes, which are supposed to be anisotropic. The new system contains 11 equations for 11 unknowns (Hunana *et al.*, 2011). Although it is much harder to simulate numerically than Hall MHD, this system is still significantly easier than a purely kinetic treatment (in phase space with six dimensions) based on the Maxwell–Vlasov equations.

---

[5] Since the Hall term leads to a modification of the dispersion relation of Alfvén waves – which is not the case with only the compressible effects in standard MHD – these modes are sometimes called kinetic Alfvén waves (Rogers *et al.*, 2001; Schekochihin *et al.*, 2009) in the sense that these new modes are originally Alfvén waves. The use of the word kinetic can be confusing since we are employing a purely fluid description; this shows that Hall MHD may be a relevant model to describe the transition from the fluid to the kinetics.

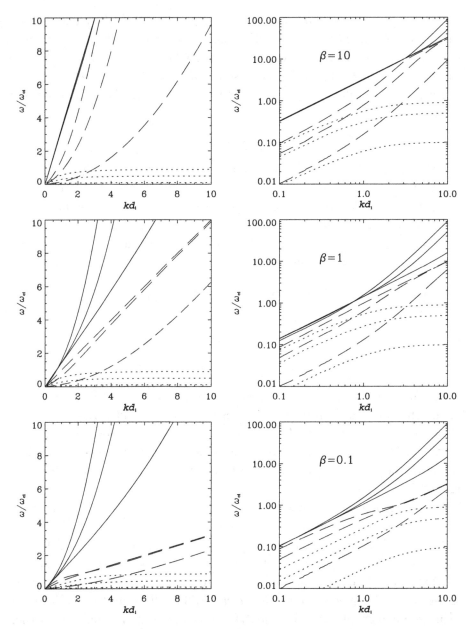

Figure 4.8 Dispersive branches (solid, dashed, and dotted lines) of compressible Hall MHD in normalized variables (see the text), in linear (left) and logarithmic (right) coordinates. The $\beta = 10$ (top), $\beta = 1$ (middle), and $\beta = 0.1$ (bottom) cases are considered, as well as the angles $\alpha = 0.9$, 0.5, and 0.1. The smaller α, the lower the branches.

5

Dynamos

The origin of the Earth's magnetic field constitutes one of the most fascinating problems of modern physics. Ever since the works of the physicist W. Gilbert in 1600, we have known that the magnetic field detected with a compass has a terrestrial origin, but a precise understanding of its production (together with the good regime of parameters) remains elusive. The generation of the Earth's magnetic field by electric currents inside our planet was proposed by Ampère just after the famous experiment done by Ørsted in 1820 (a wire carrying an electric current is able to move the needle of a compass). These currents cross the Earth's outer core, which is made of liquid metal (mainly iron) at several thousand degrees. If it were not maintained by a source these currents would disappear within several thousand years through Ohmic dissipation. Indeed, in the absence of any regenerative mechanism the Earth's magnetic field would decay in a time τ_{diff} that can be estimated with the simple relation $\tau_{\text{diff}} \sim \ell^2/\eta$. Since the Earth's metallic envelope is characterized by a thickness of ~ 2000 km and a magnetic diffusivity $\eta \sim 1$ m^2 s^{-1}, we obtain $\tau_{\text{diff}} \sim 30\,000$ years. Also, in order to explain the presence of a large-scale magnetic field on Earth since several million years ago, it is necessary to introduce the **dynamo** mechanism. It was Sir J. Larmor in 1919 who first suggested that the solar magnetic field could be maintained by what he called a self-excited dynamo, a theory explaining the formation of sunspots. Generally speaking, the dynamo effect explains the solar, stellar, and even galactic magnetic fields.

5.1 Geophysics, Astrophysics, and Experiments

5.1.1 Experimental Dynamos

The simplest experiment concerning a self-excited dynamo is Bullard's dynamo as shown in Figure 5.1: it is made of a conducting disk which rotates in a medium

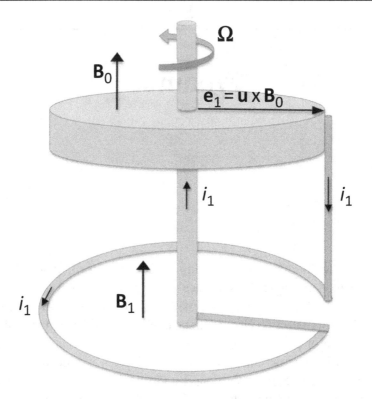

Figure 5.1 Bullard's dynamo as proposed by Bullard (1955).

where an axial magnetic field $\mathbf{B}_0$ is present. An electric wire going around the axis is connected on one side to the axis and on the other side to the disk (the electric contacts do not prevent the rotation). Because of the Lorentz force an electric potential is induced between the center and the side of the disk. The induced electromotive force $\mathbf{e}_1$ generates a current i_1 in the loop, which in turn generates an axial induced magnetic field $\mathbf{B}_1$, which adds to the initial field. Finally since the Laplace force acts against the dynamo effect, the induced magnetic field will saturate. In this experiment the direction of the rotation is fundamental: a rotation in the opposite direction to the one shown in Figure 5.1 leads to an anti-dynamo effect. In other words, we see that the topology of the currents is fundamental for the dynamo. It is easy to constrain the dynamo mechanism with such an experiment; however, we can easily imagine that it is much more difficult for a conducting fluid. In practice, we will see that there is no simple solution to the MHD dynamo.

This laboratory experiment becomes more complex if one couples it with a second disk as displayed in Figure 5.2: this is Rikitake's dynamo. In this case the induced magnetic field i_1 produced by disk 1 will act on disk 2 via its induced magnetic field and vice versa. The main new result is the introduction of chaos (see Chapter 11 for a short introduction to chaos) with possible inversions of

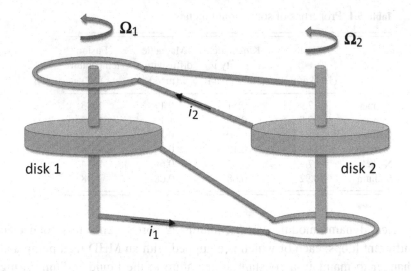

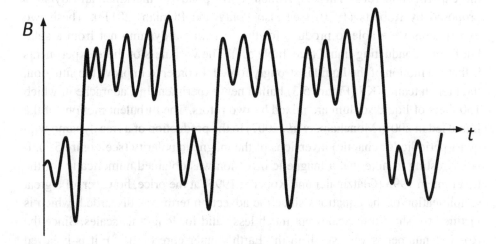

Figure 5.2 Top: dynamo proposed by Rikitake (1958). Bottom: schematic view of the temporal evolution of the (adimensionalized) magnetic field in one of the disks.

the magnetic field (which is not the case with Bullard's dynamo): just like for the Earth, we may find long periods of time without inversion and with weak variations of the magnetic field, which alternate with active periods during which inversions occur frequently.

The laboratory production of a fluid dynamo – with liquid sodium – was realized for the first time in 1999 with the Riga experiment in Latvia (Gailitis *et al.*, 2000) and the Karlsruhe experiment in Germany (Stieglitz and Müller, 2001). In the first case, a helical motion is imposed on the liquid inside a cylinder: this geometry reproduces the analytical solution of the kinematic dynamo proposed by Ponomarenko (1973). In the second case, the experiment consists essentially

Table 5.1 Properties of some liquid metals

	Density (10^3 kg/m^3)	Kinematic viscosity ν (10^{-6} m^2/s)	Magnetic diffusivity η (m^2/s)	Fusion temperature (°C)
Copper	7.9	0.51	0.17	1083
Iron	7.0	0.80	1.1	1535
Gallium	6.1	0.30	0.2	30
Mercury	13.5	0.12	0.8	−38
Nickel	7.9	0.62	0.66	1454
Sodium	0.92	0.68	0.08	98

of a cylindrical dynamo module which contains 52 vortex generators connected to three different loops, each of which is equipped with an MHD feed pump and heat exchanger to maintain a constant temperature in the liquid sodium during the experimental runs. This experiment corresponds to the kinematic dynamo proposed by Roberts (1970) (see also Ponty and Plunian (2011)). These two experiments were able to produce for the first time a dynamo not from a solid but from a conducting liquid (see Table 5.1). The weakness of these experiments is that the motion of the liquid is strongly forced. In order to improve the situation, the French team VKS (Figure 5.3) built a new experiment in Cadarache in which 150 liters of liquid sodium are mixed by two rotors. The turbulent motions of the liquid led in 2007 (Monchaux *et al.*, 2007) to the production of a free dynamo with quasi-periodic (or chaotic) inversions of the magnetic polarity (see Figure 5.3). It is interesting to note that a magnetic inversion was obtained numerically for the first time in 1995 (Glatzmaier and Roberts, 1995) at the price, however, of a great simplification of the equations since the advection term was discarded, which is justified for short time scales but much less valid for long time scales. Since the Rossby[1] number is very small in the Earth's outer core ($\sim 10^{-6}$), it is believed that the magnetic field tends to roll up (this is often called the Ω-effect). The azimuthal magnetic field component generated will eventually produce an axial component (see Figure 5.4). It is important to say, however, that the dynamics of the liquid metal is certainly much more complex than this simple picture because it is in a regime of fully developed turbulence in which (magnetostrophic) waves and coherent structures (such as vortices) interact non-linearly (see the last part of this book).

Other experiments have been developed (or are currently under development) in order to better mimic the physics of the Earth's outer core and better understand

---

[1] The Rossby number measures the relative importance of the Coriolis force compared with the non-linear term: the stronger the Coriolis force, the smaller the Rossby number. The introduction of this force in the presence of a uniform magnetic field gives rise to magnetostrophic waves whose period (in the context of the geodynamo) is of the order of thousands of years. It seems that this type of wave plays a central role in the geodynamo mechanism (see also Moffatt (1970) and Exercise II.b).

Figure 5.3 Measurements (bottom left) of intermittent magnetic field (the azimuthal component $\mathbf{B}_\theta$ is displayed) reversals made with the Von Kármán Sodium (VKS) experiment (top left) in Cadarache; credit: VKS team (see also Berhanu *et al.* (2007) and Ravelet *et al.* (2008)). The Derviche Tourneur Sodium (DTS) experiment (right) which consists of a rotating sphere of radius 21 cm subjected to an axial magnetic field parallel to the rotation axis; credit: CNRS Photothèque, H.-C. Nataf.

the particular organization of fluid flow and magnetic field. For example, the DTS experiment in Grenoble (Nataf and Gagnière, 2008) has been designed for the exploration of the magnetostrophic regime, in which the Coriolis and Laplace forces are dominant (see Figure 5.3). Unlike the previous examples, the Madison Plasma Dynamo Experiment (MPDX) does not work with liquid metals but instead works with a hot plasma. This new type of dynamo experiment is made of a spherical vacuum vessel of diameter 3 m that uses an array of powerful permanent magnets on the vessel wall to provide plasma confinement (Cooper *et al.*, 2014). The use of plasma, rather than liquid metals, to study magnetic field generation will allow one in principle to reach much higher magnetic Reynolds numbers than with liquid metal experiments and provide access to regimes closer to many astrophysical situations. It is expected that the collisional plasma dynamics will fall in the range of scales where both MHD and Hall MHD can be used.

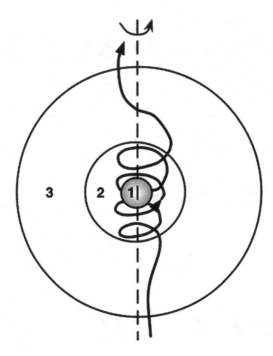

Figure 5.4 The rolling up of the Earth's magnetic field is produced by the rotation of the outer core (2) made of liquid metals (with the solid inner core in (1) and the mantle in (3)).

5.1.2 Natural Dynamos

It has been known for about 2000 years that the needle of a compass points approximately in the direction of the terrestrial North Pole. In fact, the terrestrial magnetic field has existed for millions years (Tarduno *et al.*, 2015): it forms the so-called magnetosphere. The magnetic field evolves on short time scales (of the order of hours or less) which can be observed with space probes, but also on very long time scales. Indeed, paleomagnetism tells us that the Earth's magnetic field may chaotically undergo inversion, with the last inversion having occurred 780 000 years ago (Figure 5.5). Interestingly, the origin of the reversals of Earth's magnetic field is still the subject of intense research (Pétrélis *et al.*, 2009). The understanding of the geo-dynamo requires a detailed knowledge both of the internal structure of our planet and of the solutions of the non-linear MHD equations, in which the boundary conditions play a non-negligible role. We know that the Earth consists of a solid inner core of iron and nickel with a radius of ∼1300 km and a fluid outer core of liquid iron of thickness ∼2300 km. Beyond that, we find a solid mantle of silicates and a rocky crust. The geo-dynamo originates in the fluid outer core. The turbulent motion of the liquid metal is maintained by a source of heating (the radioactive decay of elements such as

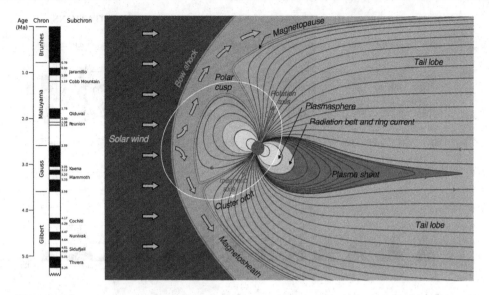

Figure 5.5 Left: temporal variation of the Earth's magnetic polarity over the last 5.25 million years: the dark zones correspond to the current polarity. Right: Earth's magnetosphere (image made by P. Robert, LPP/CNRS).

uranium and thorium). We also know that rotation probably has a strong impact on the dynamo.

Planetary magnetic fields have been detected for most of the planets of our solar system, with a range of values going from about 10^{-7} T (Mercury) to 10^{-4} T (Jupiter). The case of Mars is interesting because a remnant magnetic field is found, which is frozen into its crustal rocks. This probably means that a dynamo operated in the past. It is thought that the shut down of the dynamo is related to the evolution of the mantle convection, going from Earth-like plate tectonics to a Venus-like rigid plate. In such a situation the radial heat flux is stopped, which leads to the suppression of the convection in the core. For Jupiter and Saturn the magnetic field is strong, the Rossby number is very small ($\leq 10^{-6}$), and the dipolar magnetic axis is almost aligned with the rotation axis (for Saturn the alignment is nearly perfect; see Figure 2.3). These giant planets have extended outer cores of liquid metallic hydrogen. For Uranus and Neptune magnetic fields are also observed, with the same order of magnitude as for Saturn (10^{-5} T), but these fields are not dipolar and have a complex spatial distribution. These ice giants are thought to be composed mainly of silicate and metals in the core, water, ammonia, methane in the mantle, and a thin hydrogen–helium atmosphere.

In the case of the Sun we have information on the magnetic field via, for example, the sunspots which represent the most intense magnetic flux concentrations on the solar surface. The average magnetic field tends to increase with the area of the spot, whose diameter can exceed sometimes the size of the Earth. Since they can

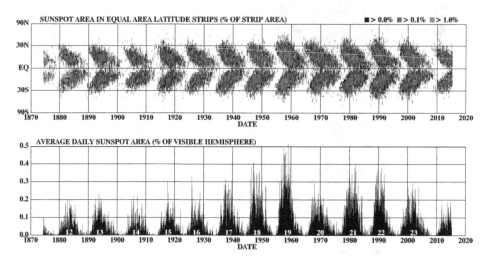

Figure 5.6 Top: migration of sunspots towards the equator. Bottom: temporal variation of the sunspot number with a periodicity of 11 years. Observations of sunspots have been obtained by the Royal Greenwich Observatory since 1874. Credits: NASA.

last for several days and up to a month it is easy to follow their motion from west to east across the solar disk. Figure 5.6 shows that a period of 11 years (called the solar cycle) appears with surprising regularity. This regularity was discovered by H. Schwabe – an amateur astronomer – who systematically recorded the appearance of sunspots during the period 1826–1851. The other remarkable aspect is that these sunspots are localized in two belts between the equator and the latitudes $\pm 30°$, and they migrate systematically towards the equator (this is often called the butterfly diagram). Then, they disappear and the Sun enters a quiet period (solar minimum) without active regions. Actually, when the polarity of the magnetic field was taken into account, it was realized that the physically relevant period of the solar cycle is 22 years, with a reversal of the large-scale magnetic field every 11 years. This evolution tells us something about the dynamo mechanism inside the Sun which tends to produce magnetic loops at moderate latitudes.

There are many other astrophysical examples where the dynamo plays a fundamental role in maintaining the magnetic field, such as stars having a convective zone (whose magnetic activity increases with the rotation rate) or galaxies (see Figure 5.7 and e.g. Brandenburg and Subramanian (2005)). The presence of a galactic magnetic field poses the problem of its growth: in fact, this means that the galactic dynamo must be efficient enough to generate a coherent field on scales as large as the galaxy. The necessary existence of an initial seed is another issue that arises for galaxies and stars.

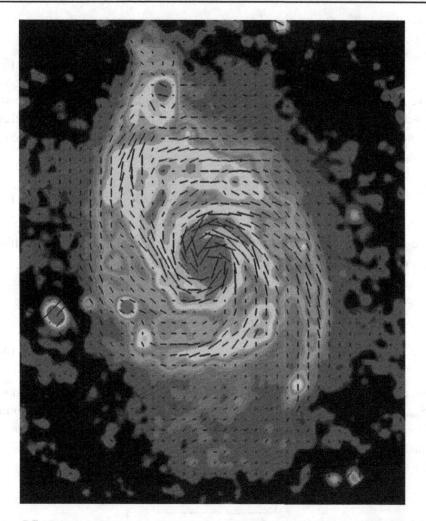

Figure 5.7 Galaxy M51: total radio emission (shading) and magnetic field orientations (segments), obtained from combined data of the VLA and Effelsberg radio telescopes at wavelength 6 cm. A coherent magnetic field at galactic scale emerges (image size: 28×34 kpc^2; distance: 9.6 Mpc; 1 pc $\simeq 3 \times 10^{16}$ m $\simeq 2 \times 10^5$ AU). Credit: Rainer Beck (MPIfR Bonn, Germany) and Andrew Fletcher (University of Newcastle, UK).

5.2   The Critical Magnetic Reynolds Number

We shall see that the magnetic energy equation gives a first constraint on the existence of the dynamo effect: the magnetic Reynolds number must be larger than a critical value.

We start with the magnetic energy Eq. (3.18) that we integrate over the volume $\mathcal{V}_\infty$ on which the Poynting vector vanishes. After the use of the standard form of Ohm's law we obtain

$$\frac{dE^B}{dt} = -\iiint (\mathbf{j} \times \mathbf{B}) \cdot \mathbf{u}\, d\mathcal{V} - \mu_0 \eta \iiint \mathbf{j}^2\, d\mathcal{V}$$

$$= \mathcal{S} - \mathcal{D}, \tag{5.1}$$

where E^B is the magnetic energy. The source of magnetic energy $\mathcal{S}$ has no sign defined whereas the Ohmic dissipation, $-\mathcal{D}$, is always negative. It is clear that a dynamo exists only if $\mathcal{S} - \mathcal{D} > 0$. We may limit these two contributions by $|\mathcal{S}|_{\max}$ and $\mathcal{D}_{\min}$. We have

$$\mathcal{D} = \frac{\eta}{\mu_0} \iiint (\nabla \times \mathbf{B})^2\, d\mathcal{V} \geq \frac{\eta}{\mu_0} \iiint \frac{\mathbf{B}^2}{L_{\max}^2}\, d\mathcal{V} = \frac{2\eta E^B}{L_{\max}^2} \equiv \mathcal{D}_{\min}, \tag{5.2}$$

where $L_{\max}$ is the maximum scale on which the magnetic field varies. Additionally, we have (with Schwartz's inequality)

$$\mathcal{S}^2 \leq u_{\max}^2 \left(\iiint (\mathbf{j} \times \mathbf{B}) d\mathcal{V} \right)^2 \leq \frac{u_{\max}^2}{\mu_0^2} \iiint \mathbf{B}^2\, d\mathcal{V} \iiint (\nabla \times \mathbf{B})^2\, d\mathcal{V}$$

$$\leq 2u_{\max}^2 E^B \left(\frac{\mathcal{D}}{\eta} \right) \equiv \mathcal{S}_{\max}^2, \tag{5.3}$$

where $u_{\max}$ is the maximum velocity. Thus, a necessary condition to get a dynamo is $|\mathcal{S}_{\max}| > \mathcal{D}_{\min}$. In the case of an integration over a ball of radius R, the maximum scale on which the magnetic field varies will be of the order of R; hence the inequality

$$\boxed{R_{\mathrm{m}} \equiv \frac{u_{\max} R}{\eta} > 1,} \tag{5.4}$$

with R_{m} the magnetic Reynolds number. Although it is approximative, this estimate reveals the existence of a critical magnetic Reynolds number below which a dynamo is impossible because the Ohmic dissipation is then greater than the stirring force. Experiments and direct numerical simulations (Nore *et al.*, 1997) have shown that this number is approximately 20 (see the discussion of the Ponomarenko dynamo, Section 5.5).

5.3 The Kinematic Regime

A simple way to tackle the dynamo problem is to consider that the velocity is given. Then, we only have to find the magnetic field without looking at its feedback on the velocity. This is the **kinematic dynamo** regime. This regime is particularly important when we investigate the initial development of the magnetic field because a small magnetic seed has a negligible effect on the velocity field.

Let us consider the induction equation of compressible standard MHD:

$$\frac{\partial \mathbf{B}}{\partial t} = \nabla \times (\mathbf{u} \times \mathbf{B}) + \eta \,\Delta \mathbf{B},$$

$$= -\mathbf{u} \cdot \nabla \mathbf{B} + \mathbf{B} \cdot \nabla \mathbf{u} - \mathbf{B}(\nabla \cdot \mathbf{u}) + \eta \,\Delta \mathbf{B}, \tag{5.5}$$

where $\mathbf{u}(\mathbf{x}, t)$ is a given velocity field (its form may be non-trivial in the case of turbulence). We have two contributions which may balance the diffusion term and amplify the magnetic field (the first term on the right-hand side is purely advective). The third term on the right-hand side is by nature compressible and contributes locally to the dynamo when the fluid is compressed: then, we have a local concentration of magnetic field lines which retain the same direction. This effect disappears in the incompressible case. The second term on the right-hand side is, however, always present: this is the main contribution to the generation (or destruction) of the magnetic field. Unlike the compressible effects, this term may amplify the magnetic field in the velocity direction (see Figure 5.8).

To illustrate the kinematic dynamo, let us consider the particular case of an axisymmetric differential rotation with initially a poloidal magnetic field:

$$\mathbf{B}(r, z, t = 0) = (B_r, 0, B_z), \tag{5.6}$$

$$\mathbf{u}(r, z, t) = (0, \Omega(z)r, 0); \tag{5.7}$$

we may think of the Sun, for which a differential rotation is observed. We obtain the equation

$$\frac{\partial \mathbf{B}}{\partial t} = \nabla \times (\mathbf{u} \times \mathbf{B}) + \eta \,\Delta \mathbf{B}$$

$$= [\partial_r(\Omega r B_r) + \partial_z(\Omega r B_z)] \, \mathbf{e}_\theta + \eta \,\Delta \mathbf{B}, \tag{5.8}$$

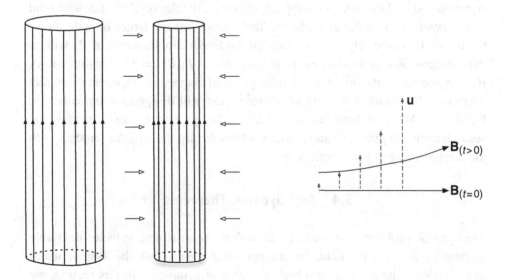

Figure 5.8 Amplification of the magnetic field by compression (left) or shearing (right).

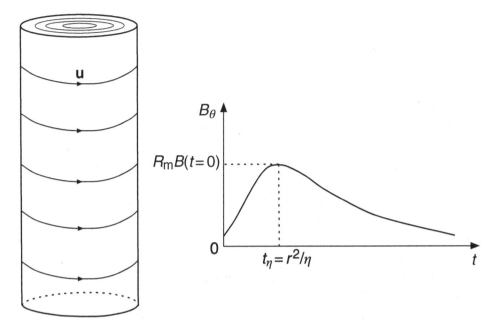

Figure 5.9 In an axisymmetric configuration, a fluid under differential rotation can generate a magnetic field only for a limited period of time.

where $\mathbf{e}_\theta$ is a toroidal unit vector. A toroidal magnetic field is thus generated by a lateral shearing of the poloidal magnetic field. Because of the form of the velocity field, the toroidal component of the magnetic field will never affect the dynamics of the poloidal component. Thus, the latter evolves following a purely diffusive equation and will eventually disappear. However, the decay of the poloidal field will depend on the value of η: the smaller η (and thus the larger R_m), the faster the decay. It is precisely in this limit that the **linear** amplification of B_θ will be the strongest. A dimensional estimate gives $B_\theta \sim R_m B(t = 0)$. This growth of B_θ is, however, limited in time: when the poloidal magnetic component is entirely destroyed, B_θ is only the result of diffusion and will disappear eventually (see Figure 5.9). What we have just described is a key process for understanding the solar dynamo. Figure 5.10 summarizes schematically the dynamo process with the strengthening of a magnetic loop.

5.4 Anti-dynamo Theorems

The dynamo problem is difficult because there is no simple velocity field able to amplify a magnetic field. In practice, that means that the system under investigation must not have too high a degree of symmetry. In this section, we shall demonstrate that there is no purely axisymmetric dynamo. This is the first

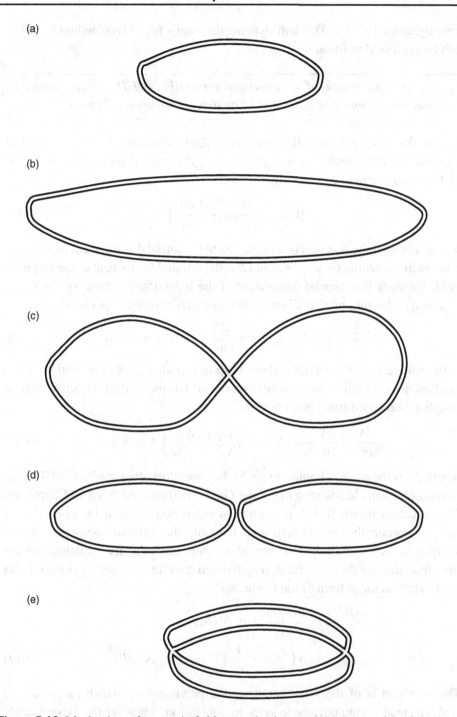

Figure 5.10 Mechanism of magnetic field strengthening: (a) initial state of the field line; (b) shearing; (c) twisting; (d) diffusion; (e) pilling.

anti-dynamo theorem. This anti-dynamo theorem – found by Cowling in 1933 – can be expressed as follows.

> *In the framework of standard incompressible MHD, an axisymmetric magnetic field cannot be sustained by an axisymmetric velocity field.*

For the demonstration we will show that, if the fields **u** and **B** are axisymmetric, then the magnetic field always decreases. In cylindrical coordinates, an axisymmetric magnetic field can be written as

$$\mathbf{B} = \left(\frac{1}{r} \frac{\partial A}{\partial z}, B, -\frac{1}{r} \frac{\partial A}{\partial r} \right),\tag{5.9}$$

where A (called the flux function) is linked to the toroidal component of the magnetic vector potential ($A \equiv -rA_\theta$) and B is the toroidal component of the magnetic field. By using the toroidal component of the induction equation for the vector potential and assuming that all the vectors are axisymmetric, we obtain

$$\frac{\partial A}{\partial t} + u_r \partial_r A + u_z \partial_z A = \frac{\partial A}{\partial t} + \mathbf{u} \cdot \nabla A = r\eta\mu_0 j_\theta.\tag{5.10}$$

Although the vector potential is defined up to a gradient, the contribution of this gradient does not affect the toroidal component. The use of the Laplacian operator in cylindrical coordinates gives finally

$$\frac{dA}{dt} \equiv \frac{\partial A}{\partial t} + \mathbf{u} \cdot \nabla A = \eta \left(\Delta - \frac{2}{r} \frac{\partial}{\partial r} \right) A = \eta \Delta^* A,\tag{5.11}$$

where Δ^* is the pseudo-Laplacian (or Stokes operator) often used for MHD equilibrium problems in tokamaks (see the Grad–Shafranov equation in Chapter 8). This equation means that A is mainly advected and diffused in time. Also, A will asymptotically tend to zero and thus only the toroidal component B will survive. In these conditions, we may show from the induction equation (we use the expression for the ∇ operator in cylindrical coordinates – see Appendix 2 – as well as the vectorial form of the Laplacian[2]):

$$\frac{d(B/r)}{dt} \equiv \frac{\partial (B/r)}{\partial t} + \mathbf{u} \cdot \nabla(B/r)$$

$$= \eta \left(\frac{1}{r} \Delta - \frac{1}{r^3} \right) B = \eta \left(\frac{1}{r^2} \Delta^*(rB) \right).\tag{5.12}$$

This equation is of the same nature as the previous one, which means that B will also tend asymptotically to zero. In conclusion, whatever the axisymmetric velocity field chosen, an axisymmetric magnetic field can only decrease in time.

---

[2] The Laplacian operator applied to a vector is different from the one applied to a scalar, whose expression (in cylindrical or spherical coordinates) is much simpler (see Appendix 2).

This theorem belongs to a family of anti-dynamo theorems which tend to demonstrate that no simple solution is possible to produce the dynamo effect. Among these theorems one can cite the statements that "A bidimensional flow cannot maintain a magnetic field" and "No magnetic field independent of the space coordinate can be maintained."[3]

5.5 The Ponomarenko Dynamo

The Ponomarenko flow – obtained by the author in 1973 – is the first example of flow that is susceptible to producing a dynamo. In this configuration, the strongly constrained velocity field is given by (we consider only the kinematic dynamo)

$$\mathbf{u} = \begin{cases} \Omega r \mathbf{e}_\theta + U \mathbf{e}_z, & r \le a, \\ \mathbf{0}, & r > a, \end{cases} \tag{5.13}$$

for an infinite and incompressible fluid. We see that the flow is axisymmetric (hence the use of the cylindrical coordinates). The velocity field corresponds to a helical vortex (see Figure 5.11), i.e. to a structure whose kinetic helicity is non-zero. Consequently, the magnetic field maintained cannot be axisymmetric too because of the first anti-dynamo theorem. The solution of the problem is not easy to derive and we need to introduce, as we will see, the Bessel functions.

Since the velocity is given, we need only consider the induction equation

$$\frac{\partial \mathbf{B}}{\partial t} = \nabla \times (\mathbf{u} \times \mathbf{B}) + \eta \Delta \mathbf{B}, \tag{5.14}$$

for which we shall find solutions of the form

$$\mathbf{B}(r, \theta, z, t) = \mathbf{B}(r)\exp(i(m\theta + kz) + \gamma t). \tag{5.15}$$

Our goal is to check that it is possible to get a positive growth rate, i.e. $\text{Re}(\gamma) > 0$ (it is fundamentally a study of instability, to which we will return in the next part of the book). The projections of the induction equation on the r and θ directions give, respectively,[4]

$$\gamma B_r = -i(m\Omega + kU)B_r$$
$$+ \eta \left(\frac{d^2 B_r}{dr^2} + \frac{1}{r}\frac{dB_r}{dr} - \frac{(m^2 + k^2 r^2 + 1)}{r^2}B_r - \frac{2im}{r^2}B_\theta \right), \tag{5.16}$$

$$\gamma B_\theta = -i(m\Omega + kU)B_\theta$$
$$+ \eta \left(\frac{d^2 B_\theta}{dr^2} + \frac{1}{r}\frac{dB_\theta}{dr} - \frac{(m^2 + k^2 r^2 + 1)}{r^2}B_\theta + \frac{2im}{r^2}B_r \right) + r\frac{d\Omega}{dr}B_r. \tag{5.17}$$

---

[3] Analytical works have shown that a variation of the conductivity may modify these conclusions somewhat and produce a dynamo effect with, for example, a uniform flow between two plates (Busse and Wicht, 1992).

[4] Do not forget to express the ∇ operator correctly in cylindrical coordinates (see Appendix 2).

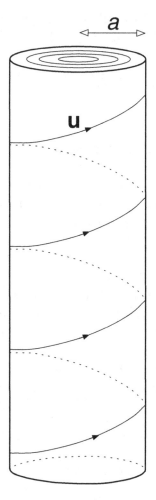

Figure 5.11 Helical structure favorable to the amplification of a magnetic field: only one velocity field line is plotted at the surface of the cylinder of radius $r = a$. Note that it is this type of structure that is naturally produced in turbulence.

In this problem Ω is constant but its derivative is non-zero in $r = a$; this information will be used later. It is not necessary to project on z since the zero-divergence condition for the magnetic field can give us this expression. By introducing the notations

$$B_{\pm} = B_r \pm iB_{\theta} \,, \tag{5.18}$$

$$x = \frac{r}{a} \,, \tag{5.19}$$

$$\tau_{\eta} = \frac{a^2}{\eta} \,, \tag{5.20}$$

$$q^2 = k^2 a^2 + \gamma \tau_{\eta} + i(m\Omega + kU)\tau_{\eta} \,, \tag{5.21}$$

$$s^2 = k^2 a^2 + \gamma \tau_\eta \,, \tag{5.22}$$

we obtain for $x \leq 1$

$$x^2 \frac{d^2 B_\pm}{dx^2} + x \frac{dB_\pm}{dx} - \left((m \pm 1)^2 + q^2 x^2\right) B_\pm = 0 \,, \tag{5.23}$$

and for $x > 1$

$$x^2 \frac{d^2 B_\pm}{dx^2} + x \frac{dB_\pm}{dx} - \left((m \pm 1)^2 + s^2 x^2\right) B_\pm = 0 \,. \tag{5.24}$$

We recognize the modified Bessel function of order $m \pm 1$ (the mathematical properties are recalled in Chapter 10) whose solutions are

$$B_\pm(x) = \begin{cases} C_\pm \frac{I_{m\pm1}(qx)}{I_{m\pm1}(q)}, & x \leq 1, \\ \tilde{C}_\pm \frac{K_{m\pm1}(sx)}{K_{m\pm1}(s)}, & x > 1. \end{cases} \tag{5.25}$$

We use the boundary conditions as a constraint for the solutions. For example, by imposing the continuity of $B_\pm$ at $x = 1$, we trivially obtain

$$C_\pm = \tilde{C}_\pm \,. \tag{5.26}$$

A second constraint is obtained by integrating the equations around $r = a$. From expression (5.17), we obtain[5]

$$0 = \lim_{\varepsilon \to 0} \left[\eta \int_{a-\varepsilon}^{a+\varepsilon} \frac{d^2 B_\theta}{dr^2} \, dr + \int_{a-\varepsilon}^{a+\varepsilon} r \frac{d\Omega}{dr} B_r \, dr \right] \,, \tag{5.27}$$

which gives

$$a\Omega B_r = \eta \left[\frac{dB_\theta}{dr} \right]_{a-\varepsilon}^{a+\varepsilon} \,. \tag{5.28}$$

This expression leads finally to the boundary condition

$$\pm i\Omega \tau_\eta \left(\frac{B_+ + B_-}{2} \right) = \left[\frac{dB_\pm}{dx} \right]_{x=1_-}^{x=1_+} \,. \tag{5.29}$$

The combination of the solutions (5.25) with the boundary conditions (5.26) and (5.27) gives the dispersive relation

$$G_+ G_- = i\Omega \tau_\eta \left(\frac{G_+ - G_-}{2} \right) \,, \tag{5.30}$$

with

$$G_\pm \equiv q \frac{I'_{m\pm1}(q)}{I_{m\pm1}(q)} - s \frac{K'_{m\pm1}(s)}{K_{m\pm1}(s)} \,, \tag{5.31}$$

---

[5] Here we see why it was relevant to keep the term $r(d\Omega/dr)B_r$ in Eq. (5.17).

where $'$ is the derivative sign. This dispersive relation can give us the necessary condition for a dynamo, i.e. the condition corresponding to $\mathrm{Re}(\gamma) > 0$. In the most general case it is impossible to get this condition analytically; however, a numerical estimate is possible. In the small-scale limit (i.e. for $ka \gg 1$) or for a small magnetic resistivity ($\Omega \tau_\eta \gg 1$), it is possible to derive a simple expression for this condition. In these limits, we obtain asymptotically

$$\sqrt{2z\pi}\, I_m(z) \simeq \exp(z)\left(1 - \frac{4m^2 - 1}{8z}\right), \tag{5.32}$$

$$\sqrt{\frac{2z}{\pi}}\, K_m(z) \simeq \exp(-z)\left(1 + \frac{4m^2 - 1}{8z}\right), \tag{5.33}$$

where $|\arg(z)| < \pi/2$. Then, we obtain the following estimate:

$$G_\pm \simeq q + s + (m^2/2 \pm m + 3/8)(1/q + 1/s), \tag{5.34}$$

which allows us to simplify the dispersion relation:

$$(q + s)qs = im\Omega\tau_\eta . \tag{5.35}$$

If we neglect the purely imaginary part of q, we find

$$\gamma\tau_\eta \simeq \left(\frac{m\Omega\tau_\eta}{2}\right)^{2/3} \exp(i\pi/3) - k^2 a^2 , \tag{5.36}$$

which means, in particular, that the solution implies oscillations. A dynamo may be obtained if

$$\mathrm{Re}(\gamma) > 0, \tag{5.37}$$

and thus if

$$\Omega\tau_\eta > \frac{4\sqrt{2}(ka)^3}{m} . \tag{5.38}$$

Clearly, we see that the dynamo effect occurs more easily when the fluid rotates rapidly and/or when the magnetic resistivity is low. This condition may also be written with the magnetic Reynolds number, which is defined as

$$R_\mathrm{m} = \frac{\tau_\eta}{\tau_\mathrm{nl}}, \tag{5.39}$$

where the typical non-linear time for this problem is

$$\tau_\mathrm{nl} = \frac{a}{\sqrt{\Omega^2 a^2 + U^2}} . \tag{5.40}$$

Thus, the condition to get the Ponomarenko dynamo is

$$\boxed{R_\mathrm{m} > 4\sqrt{2}\sqrt{1 + \left(\frac{U}{\Omega a}\right)^2}\, \frac{(ka)^3}{m} .} \tag{5.41}$$

When $m = 0$ the condition cannot be satisfied: in this case, we have an axisymmetric magnetic field which cannot be maintained because of the first anti-dynamo theorem. In the general case, numerical simulations show that the critical magnetic Reynolds number is around 20: this dynamo corresponds to a judicious choice of m and k which minimizes the values of Ω and U. As a result, the higher the Reynolds number, the more we can excite modes at small scales.

5.6 The Turbulent Dynamo

We have seen that to produce a dynamo it is better to have a velocity field with a weak symmetry. Then, we can easily be convinced that if this field is turbulent then the chance of getting a dynamo is high. It turns out that the natural dynamos discussed above are all turbulent. The last part of this book will be devoted to turbulence; without being specialists in this subject, we nevertheless address from now on a few concepts that can be used to understand some properties of the turbulent dynamo.

5.6.1 Kinematic Mean Field Theory

In order to study the role of turbulence, we shall decompose the fields into a mean part (with slow variations in space and time) and a fluctuating part (with fast variations in space and time):

$$\mathbf{b} = \mathbf{B}_0 + \mathbf{b}', \tag{5.42}$$

$$\mathbf{u} = \mathbf{U}_0 + \mathbf{u}', \tag{5.43}$$

where $\mathbf{U}_0 \equiv \langle \mathbf{u} \rangle$ and $\mathbf{B}_0 \equiv \langle \mathbf{b} \rangle$ are mean quantities. (The meaning of $\langle \cdot \rangle$ will be discussed in detail in Chapter 11.) The fluctuations, being turbulent by nature, will be supposed to be weak in amplitude. The introduction of these decompositions into the averaged induction equation gives

$$\frac{\partial \mathbf{B}_0}{\partial t} = \nabla \times [\mathbf{U}_0 \times \mathbf{B}_0] + \nabla \times \langle \mathbf{u}' \times \mathbf{b}' \rangle + \eta \, \Delta \mathbf{B}_0 . \tag{5.44}$$

We see that the small-scale turbulence plays a role in the evolution of the mean magnetic field through the second term on the right-hand side. The equation for the fluctuating part gives

$$\frac{\partial \mathbf{b}'}{\partial t} = \nabla \times \left[\mathbf{U}_0 \times \mathbf{b}' + \mathbf{u}' \times \mathbf{B}_0 \right] + \nabla \times \left[\mathbf{u}' \times \mathbf{b}' - \langle \mathbf{u}' \times \mathbf{b}' \rangle \right] + \eta \, \Delta \mathbf{b}' . \tag{5.45}$$

This equation is linear in $\mathbf{b}'$ (we suppose that the velocity is independent of the magnetic field). In particular, we see that the second term on the right-hand side acts like a source that is independent of the magnetic fluctuations.

Let us assume that initially $\mathbf{b}' = 0$. The linearity in $\mathbf{b}'$ of Eq. (5.45) tells us that the magnetic fluctuations are initially linearly dependent on $\mathbf{B}_0$. Consequently, that is also the case for the term $\langle \mathbf{u}' \times \mathbf{b}' \rangle$ in Eq. (5.44). This remark suggests that we should model this term as

$$\langle \mathbf{u}' \times \mathbf{b}' \rangle_i \simeq \alpha_{ij} B_{0j}, \tag{5.46}$$

where α_{ij} is an unknown tensor. In the case of a homogeneous and isotropic turbulence, we have

$$\alpha_{ij} = \alpha \delta_{ij}, \tag{5.47}$$

which allows us to rewrite Eq. (5.44) as

$$\boxed{\frac{\partial \mathbf{B}_0}{\partial t} = \nabla \times [\mathbf{U}_0 \times \mathbf{B}_0] + \alpha \, \nabla \times \mathbf{B}_0 + \eta \, \Delta \mathbf{B}_0 \,.} \tag{5.48}$$

The first-order effect of turbulence is thus to introduce a new term in the mean induction equation: this is called the α-**effect**.[6] As we will see below, the α-effect allows the generation of a magnetic field without taking into account its feedback: it is basically the kinematic dynamo regime.

5.6.2   The α-effect

The α-effect is important because it explains how small-scale turbulence can efficiently produce a magnetic field. A simple illustration can be given by considering the case of a force-free field (i.e. the situation where the Laplace force does not play a role; see Chapter 8). We shall assume that the mean velocity is null, which gives

$$\frac{\partial \mathbf{B}_0}{\partial t} = \alpha \, \nabla \times \mathbf{B}_0 + \eta \, \Delta \mathbf{B}_0 \,. \tag{5.49}$$

With

$$\mathbf{B}_0 = \tilde{\mathbf{B}}_0(\mathbf{r}) \exp(\gamma t), \tag{5.50}$$

and $\nabla \times \tilde{\mathbf{B}}_0 = K \tilde{\mathbf{B}}_0$ (the force-free field assumption), we obtain the dispersion relation

$$\gamma = \alpha K - \eta K^2, \tag{5.51}$$

which demonstrates that if $\alpha > \eta K$ the magnetic field is amplified.

---

[6] The origin of this name is directly linked to the notation used in the article published in 1966 (Steenbeck *et al.*, 1966).

5.7 Conclusion

To summarize the situation, we may say that the dynamo effect is composed of two regimes: the first one is the kinematic regime which was introduced through the α-effect. The second one is the non-linear regime (which is often turbulent), for which the fluctuations of the magnetic field can no longer be considered small anymore. A saturation of the magnetic field amplitude is then observed (like in Bullard's dynamo as discussed above). This second regime is much more difficult to analyze and numerical simulations are necessary (see Figure 5.12).

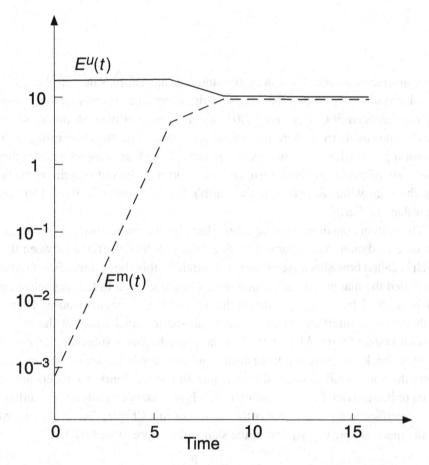

Figure 5.12 Schematic view of the exponential increase and saturation of the magnetic energy in a turbulent dynamo. In the stationary regime an equipartition between the kinetic (solid line) and magnetic (dashed line) energies is found. For an illustration with direct numerical simulations, see e.g. Mininni *et al.* (2005).

6

Discontinuities and Shocks

The Universe is a great laboratory for studying natural plasmas. In the case of the solar system, the Sun is the source of the interplanetary plasma that spreads at a rate between 300 km/s and 1000 km/s. This plasma may encounter several obstacles during its trip: asteroids, comets, or planets. The most interesting obstacles for a physicist are the magnetized planets. With their magnetosphere, these planets significantly increase their cross-section and therefore their interaction with the solar wind; for example the Earth's magnetosphere is about 150 times larger than the Earth.

The system constituted by the solar wind plus the magnetosphere is naturally in a state of dynamic equilibrium, with a relatively thin interface between them which is called bow shock (see Figure 6.1). Behind this shock, there is a turbulent area called the magnetosheath which serves as a transition to the magnetosphere that is reached by crossing a discontinuity called the magnetopause. There is another type of interface for the solar wind: the terminal shock at the edge of the solar system ($\sim$100 AU) when the wind speed becomes subsonic. Beyond the terminal shock, we have the heliosheath and then the heliopause[1] (the interface where the solar wind is stopped by the interstellar medium). To understand the nature of these shocks and discontinuities, it is necessary to study the evolution of a thin interface in a plasma; that is the subject of this Chapter. To do this, we will use the macroscopic description of the standard compressible MHD.

6.1 Rankine–Hugoniot Conditions

The method generally used to get the conditions of a plasma around a discontinuity is to integrate the conservation laws – that we established in

---

[1] The probes Voyager 1 and 2 launched in 1977 have crossed the terminal shock and should soon reach the heliopause.

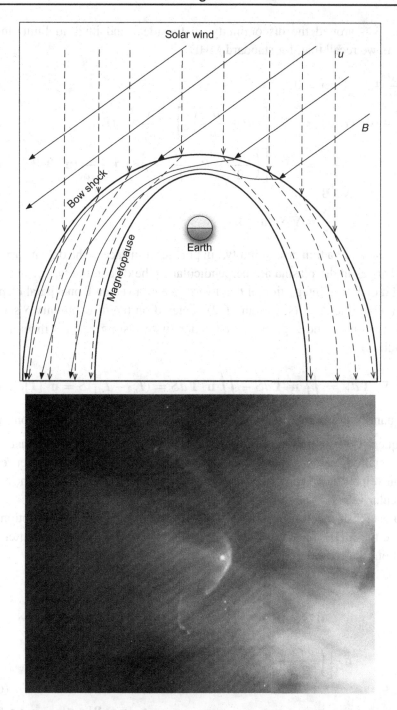

Figure 6.1 Top: solar-wind–Earth interfaces with a bow shock and a discontinuity called the magnetopause (the bow shock is essentially limited to the day side). Bottom: bow shock in the Orion Nebula; credits NASA/HST.

Chapter 3 – around the discontinuity, in the ideal and inviscid limit. In this situation, we recall that, for standard MHD,

$$\frac{\partial \rho}{\partial t} + \nabla \cdot (\rho \mathbf{u}) = 0 \,, \tag{6.1}$$

$$\frac{\partial \rho \mathbf{u}}{\partial t} + \nabla \cdot (\rho \mathbf{u}\mathbf{u}) = -\nabla \cdot \left[\left(P + \frac{B^2}{2\mu_0} \right) \mathbf{I} \right] + \frac{1}{\mu_0} \nabla \cdot (\mathbf{B}\mathbf{B}) \,, \tag{6.2}$$

$$\frac{\partial E}{\partial t} = -\nabla \cdot \left[\left(\frac{\rho u^2}{2} + P + \frac{B^2}{\mu_0} + \rho e \right) \mathbf{u} - \frac{1}{\mu_0}(\mathbf{u} \cdot \mathbf{B})\mathbf{B} \right], \tag{6.3}$$

$$\nabla \cdot \mathbf{B} = 0 \,, \tag{6.4}$$

$$\frac{\partial \mathbf{B}}{\partial t} = \nabla \times (\mathbf{u} \times \mathbf{B}) \,. \tag{6.5}$$

In the case of a thin discontinuity,[2] in practice a surface S, the only measurable local changes in the plasma are perpendicular to the discontinuity, i.e. along the normal $\mathbf{n}$ of S. The integration of the divergence of a vector $\mathbf{T}$ on a parallelepiped of negligible thickness (see Figure 6.2) centered on the discontinuity S separating the regions 1 and 2 gives at first order (with Ostrogradsky's theorem; see Appendix 2)

$$\iiint \nabla \cdot \mathbf{T} \, d\mathcal{V} = \iint_2 \mathbf{n} \cdot \mathbf{T} \, d\mathcal{S} - \iint_1 \mathbf{n} \cdot \mathbf{T} \, d\mathcal{S} = (T_{n2} - T_{n1})\mathcal{S} = \mathbf{n} \cdot [\mathbf{T}]\mathcal{S} \,. \tag{6.6}$$

The quantity $[\mathbf{T}]$ is the jump of $\mathbf{T}$ when we cross the surface of discontinuity.

Subsequently, we will assume for simplicity that the surface $\mathcal{S}$ is stationary. The MHD equations being invariant by Galilean transformation (in the presence of a uniform velocity), we can generalize the study to the case of a surface moving at a particular velocity.

To establish the **jump conditions**, we use the stationarity assumption and integrate the MHD equations over the previous volume. One obtains (after simplification) the conditions

$$\mathbf{n} \cdot [\rho \mathbf{u}] = 0 \,, \tag{6.7}$$

$$\mathbf{n} \cdot [\rho \mathbf{u}\mathbf{u}] + \mathbf{n} \cdot \left[P + \frac{B^2}{2\mu_0} \right] - \frac{1}{\mu_0}\mathbf{n} \cdot [\mathbf{B}\mathbf{B}] = 0 \,, \tag{6.8}$$

$$\mathbf{n} \cdot \left[\left(\frac{\rho u^2}{2} + P + \frac{B^2}{\mu_0} + \rho e \right) \mathbf{u} - \frac{1}{\mu_0}(\mathbf{u} \cdot \mathbf{B})\mathbf{B} \right] = 0 \,, \tag{6.9}$$

$$\mathbf{n} \cdot [\mathbf{B}] = 0 \,, \tag{6.10}$$

$$\mathbf{t} \cdot [\mathbf{u} \times \mathbf{B}] = 0 \,. \tag{6.11}$$

---

[2] To be more precise, we assume that the thickness of the discontinuity is significantly smaller than the scale of variation of the MHD fluid, but also significantly larger than the Debye length and the ion Larmor radius (Baumjohann and Treumann, 1996).

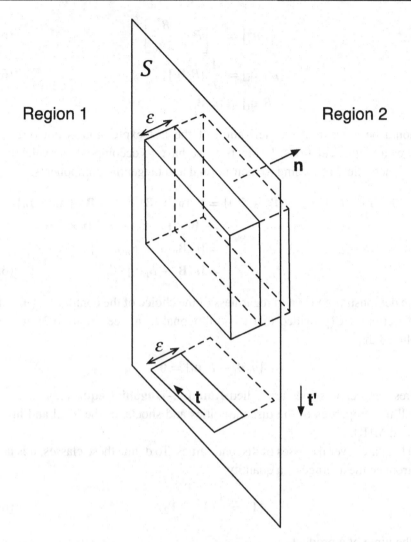

Figure 6.2 A parallelepipedal volume (and surface) crossing the discontinuity S separating regions 1 and 2 characterized by a thickness (width) that is negligible and of the order of ε and a lateral surface (length) small enough for one to consider that the properties of the plasma remain unchanged.

Note that, to get condition (6.11), we have to integrate Eq. (6.5) not over a volume but over a surface cutting the discontinuity and must use Stokes' theorem (see Appendix 2). The orientation of the sides of this surface is given by the unit vector $\mathbf{t}$ (see Figure 6.2).

One sees immediately that condition (6.10) involves the continuity of the normal component of the magnetic field $[B_n] = 0$. Similarly, the condition (6.7) means that the normal component of the momentum is continuous, i.e. $[\rho u_n] = 0$. Conditions (6.8) and (6.11) can be written in a simpler way by introducing the transverse components and using the previous properties. This yields

$$\left[\rho u_n^2\right] = -\left[P + \frac{B^2}{2\mu_0}\right], \tag{6.12}$$

$$\left[\rho u_n \mathbf{u_t}\right] = \frac{1}{\mu_0}[B_n \mathbf{B_t}], \tag{6.13}$$

$$[B_n \mathbf{u_t}] = [u_n \mathbf{B_t}]. \tag{6.14}$$

It should be noted that the derivation of the last expression is not direct and requires a few calculations. In practice, we have to decompose the velocity and the magnetic field according to their normal and tangential components:

$$\begin{aligned} 0 = \mathbf{t} \cdot [(\mathbf{u_n} + \mathbf{u_t}) \times (\mathbf{B_n} + \mathbf{B_t})] &= \mathbf{t} \cdot [\mathbf{u_n} \times \mathbf{B_t} + \mathbf{u_t} \times \mathbf{B_n} + \mathbf{u_t} \times \mathbf{B_t}] \\ &= [\mathbf{B_t} \cdot (\mathbf{t} \times \mathbf{u_n}) - \mathbf{u_t} \cdot (\mathbf{t} \times \mathbf{B_n})] \\ &= [u_n \mathbf{B_t} \cdot \mathbf{t'} - B_n \mathbf{u_t} \cdot \mathbf{t'}] \\ &= [u_n \mathbf{B_t} - B_n \mathbf{u_t}] \cdot \mathbf{t'}. \end{aligned} \tag{6.15}$$

As the demonstration is valid regardless of the choice of the contour and therefore the direction of $\mathbf{t}$ (provided that $\mathbf{t}$ is orthogonal to $\mathbf{n}$; see Figure 6.2), it can be concluded that

$$[u_n \mathbf{B_t} - B_n \mathbf{u_t}] = \mathbf{0}. \tag{6.16}$$

The resulting expressions are called **Rankine–Hugoniot equations**. They contain all the properties of the discontinuities and shocks of the ideal and inviscid standard MHD.

There are several classes of discontinuities. To define these classes, it is useful to introduce the average of a quantity,

$$\langle \mathbf{T} \rangle = \frac{1}{2}(\mathbf{T}_1 + \mathbf{T}_2), \tag{6.17}$$

and the jump of a product,

$$\left[\mathbf{T} \cdot \tilde{\mathbf{T}}\right] = [\mathbf{T}] \cdot \langle \tilde{\mathbf{T}} \rangle + \langle \mathbf{T} \rangle \cdot [\tilde{\mathbf{T}}]. \tag{6.18}$$

By introducing $F \equiv \rho u_n$ and $\mathcal{V} \equiv 1/\rho$, one obtains the linear system

$$F[\mathcal{V}] - [u_n] = 0, \tag{6.19}$$

$$F[\mathbf{u}] + \mathbf{n}[P] + \frac{1}{\mu_0}\mathbf{n}\left(\langle \mathbf{B} \rangle \cdot [\mathbf{B}]\right) - \frac{1}{\mu_0}B_n[\mathbf{B}] = 0, \tag{6.20}$$

$$F\langle \mathcal{V} \rangle[\mathbf{B}] + \langle \mathbf{B} \rangle[u_n] - B_n[\mathbf{u}] = 0, \tag{6.21}$$

$$[B_n] = 0, \tag{6.22}$$

$$F\left([e] + \langle P \rangle[\mathcal{V}] + \frac{1}{4\mu_0}[\mathbf{B_t}]^2[\mathcal{V}]\right) = 0. \tag{6.23}$$

Considering jumps as the unknowns and the averages as known, we have a system of eight equations (the projection of (6.21) on the normal direction gives a trivial relation) for nine unknowns ($[\mathcal{V}]$, $[u_n]$, $[\mathbf{u}_t]$, $[P]$, $[B_n]$, $[\mathbf{B}_t]$, and $[e]$). To close this system we use a polytropic closure. Solutions exist if and only if the determinant of the matrix is zero. This condition gives us the following equation (of seventh order in F):[3]

$$F\left(F^2 - \frac{B_n^2}{\mu_0 \langle \mathcal{V} \rangle}\right)\left(F^4 + F^2\left(\frac{[P]}{[\mathcal{V}]} - \frac{\langle \mathbf{B} \rangle^2}{\mu_0 \langle \mathcal{V} \rangle}\right) - \frac{B_n^2}{\mu_0 \langle \mathcal{V} \rangle}\frac{[P]}{[\mathcal{V}]}\right) = 0. \quad (6.24)$$

The derivation of this equation is easier with e.g. Maple and noticing that

$$[e] = \frac{1}{\gamma - 1}\left[\frac{P}{\rho}\right] = \frac{1}{\gamma - 1}\left(\langle \mathcal{V} \rangle [P] + \langle P \rangle [\mathcal{V}]\right). \quad (6.25)$$

6.2 Discontinuities

There are three classes of discontinuities. The first class is determined by the condition

$$F_{\mathrm{I}} = 0, \quad (6.26)$$

which corresponds to a zero mass flux across the discontinuity. We will see that this class includes the **tangential** and **contact** discontinuities. The second class corresponds to the condition

$$F_{\mathrm{II}} = \pm\frac{B_n}{\sqrt{\mu_0 \langle \mathcal{V} \rangle}}, \quad (6.27)$$

and hence to a non-zero value of the mass flux. We associate with this class the **rotation** discontinuity. To define the third class, we have to cancel out the expression in the last parenthesis of (6.24), which gives the following quartic equation:

$$F_{\mathrm{III}}^4 + F_{\mathrm{III}}^2\left(\frac{[P]}{[\mathcal{V}]} - \frac{\langle \mathbf{B} \rangle^2}{\mu_0 \langle \mathcal{V} \rangle}\right) - \frac{B_n^2}{\mu_0 \langle \mathcal{V} \rangle}\frac{[P]}{[\mathcal{V}]} = 0. \quad (6.28)$$

We see that the solutions depend on the jumps in pressure and volume; they can be written as

$$F_{\mathrm{III}}^2 = -\frac{1}{2}\left(\frac{[P]}{[\mathcal{V}]} - \frac{\langle \mathbf{B} \rangle^2}{\mu_0 \langle \mathcal{V} \rangle}\right) \pm \sqrt{\frac{\Delta}{4}}, \quad (6.29)$$

---

[3] The expression for the determinant in its original form does not show the pressure and volume jumps. The use of relation (6.25) can significantly simplify the writing of this expression but introduces jumps.

with

$$\Delta = \left(\frac{[P]}{[\mathcal{V}]} - \frac{\langle \mathbf{B} \rangle^2}{\mu_0 \langle \mathcal{V} \rangle} \right)^2 + \frac{4B_n^2}{\mu_0 \langle \mathcal{V} \rangle} \frac{[P]}{[\mathcal{V}]}. \tag{6.30}$$

These solutions are **shocks** (or shock waves).

6.2.1   Tangential and Contact Discontinuities

We have seen that the first family of discontinuities is characterized by a zero mass flux. From conditions (6.19)–(6.23), we can deduce that

$$[u_n] = 0, \tag{6.31}$$

$$[\mathbf{u}_t] = 0, \tag{6.32}$$

$$[B_n] = 0, \tag{6.33}$$

$$[\mathbf{B}_t] = 0, \tag{6.34}$$

$$[P] = 0. \tag{6.35}$$

Only the mass density $[\mathcal{V}]$ varies during the crossing of a discontinuity: it is a **contact** discontinuity. Since the pressure remains unchanged, a density variation implies a change in temperature that is in practice quickly erased by the creation of a heat flux. Therefore such a discontinuity cannot persist for very long.

The second – and most interesting because less trivial – solution corresponds to a purely tangential magnetic field. In this case, we have the conditions

$$[u_n] = 0, \tag{6.36}$$

$$B_n = 0, \tag{6.37}$$

$$\left[P + \frac{\mathbf{B}_t^2}{2\mu_0} \right] = 0. \tag{6.38}$$

The last of these conditions implies the continuity of the total pressure; it is a **tangential** discontinuity. As in the previous case, the mass density may vary. The best-known example of tangential discontinuity is (in a first approximation) the magnetopause, which is the thin interface between the magnetosheath (subsonic medium) and the magnetosphere (see Figure 6.3).

6.2.2   Rotational Discontinuity

We assume now that the mass flux is non-zero and in addition that the density remains constant (which simplifies the analysis). Under these conditions, we get $[\mathcal{V}] = 0$ and thus the jump conditions

$$[u_n] = 0, \tag{6.39}$$

$$[P] = 0, \tag{6.40}$$

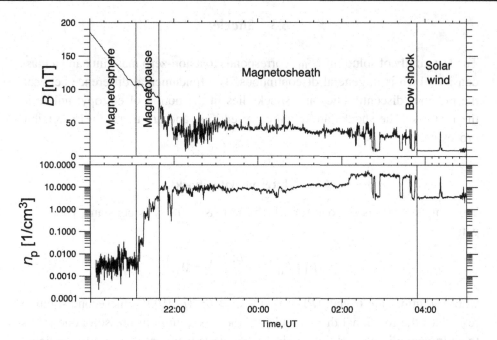

Figure 6.3 The different interfaces of the magnetosphere crossed by the Cluster4/ESA probe (February 2–3, 2003): variations in universal time of the magnetic field modulus (top) and proton density (bottom). The FGM and CIS data come from AMDA (CDPP, IRAP, France).

$$[B^2] = 0, \tag{6.41}$$

$$[B_n] = 0, \tag{6.42}$$

$$\left[\mathbf{u}_t - \frac{\mathbf{B}_t}{\sqrt{\rho\mu_0}}\right] = 0. \tag{6.43}$$

It is a **rotational** discontinuity since the last expression shows that, if the tangential component of the velocity turns during the crossing of the discontinuity, the magnetic field also turns because these two vectors are parallel. In particular, we see that the tangential component of the velocity jump is exactly equal to the tangential component of the Alfvén speed jump. To these jump conditions, one must add the relationship on the amplitude of the normal components (Walen's relation), namely

$$u_n = \frac{B_n}{\sqrt{\rho\mu_0}}. \tag{6.44}$$

Such rotational discontinuity sometimes occurs locally in the magnetopause during the phenomenon of magnetic reconnection: indeed, the magnetic topology changes and material can then penetrate into the magnetosphere. In this case, we have a non-zero normal flux of matter (F_{II}) and the tangential discontinuity is replaced temporarily by a rotational discontinuity.

6.3 Shocks

The last family of solutions (F_{III}) corresponds to a non-zero mass flux and a mass density which is in general discontinuous. The fundamental difference between the previous discontinuities and shocks lies in the notion of entropy: unlike in the first case, the shocks are a source of entropy and are therefore irreversible processes.

6.3.1 Intermediate Shocks

The simplest case is the one for which the mass density is constant. Then, one gets

$$[P]\left(F_{\mathrm{III}}^2 - \frac{B_{\mathrm{n}}^2}{\mu_0\langle\mathcal{V}\rangle}\right) = 0\,, \tag{6.45}$$

whose solutions are trivial: $[P] = 0$ or $F_{\mathrm{III}}^2 = B_{\mathrm{n}}^2/(\mu_0\langle\mathcal{V}\rangle)$. This second solution resembles the rotational discontinuity. If there is a jump in pressure, one refers to an **intermediate** shock; otherwise (when there is no pressure jump) we find a rotational discontinuity.

6.3.2 True Shock

In all other cases, Eq. (6.28) has two pairs of solutions combined with jumps in pressure and density. Among the possible solutions, only those corresponding to $F_{\mathrm{III}}^2 > 0$ are physically relevant. The best-known natural shock is the bow shock (see Figure 6.1) near the Earth. This shock reflects the interaction between the supersonic solar wind and the magnetospheric region. In addition, we make the distinction between perpendicular (the normal vector to the shock is perpendicular to the magnetic field of the solar wind) and parallel (the normal vector is parallel to the magnetic field) shocks.

6.4 Collisionless Shocks

To conclude, it is worth recalling that the Rankine–Hugoniot conditions are valid as long as the thickness of the discontinuity is significantly smaller than the scale of variation of the MHD fluid, and is significantly larger than the Debye length and the ion Larmor radius. However, it sometimes happens that the complex processes of dissipation within the discontinuity affect the behavior of the plasma outside the discontinuity. In this case, the Rankine–Hugoniot conditions are only a rough approximation of the physics and kinetic effects must be taken into account in the treatment of the discontinuity. For example, the increase of entropy in the case of shocks is associated with the transformation of a part of the kinetic energy into heat. If collisions are sufficiently numerous, the source of entropy can be associated with collisions. Otherwise, one speaks of collisionless shock and the entropy is due to collisionless processes like the Landau damping. The bow shock in the neighbourhood of the Earth is an example of a collisionless shock. In astrophysics, the phenomenology of collisionless shocks is associated with particle acceleration, radio-wave emissions, and turbulence (Belmont *et al.*, 2013).

7

Magnetic Reconnection

Magnetic reconnection is a fundamental process in plasma physics that allows the transfer of energy from the magnetic field to the plasma in the form of kinetic energy, thermal energy, or particle acceleration (Yamada *et al.*, 2010). The basic process of magnetic reconnection is the following: when two magnetic field lines of opposite directions are close enough, an intense current sheet is created between the two and a topological reorganization of the magnetic field lines occurs. Basically, this mechanism involves a violation of Alfvén's theorem whose origin is the magnetic diffusivity in standard MHD, or for example the Hall effect in collisionless plasmas. In this chapter, we present the elementary mechanism of magnetic reconnection whose main applications range from solar flares – the most violent events in the solar system – to magnetic substorms in planetary magnetospheres which produce spectacular aurorae (see Figure 7.1). Magnetic reconnection is also invoked for stellar coronae, accretion disks, dynamos, and tokamaks, and laboratory experiments have been designed specifically to study this phenomenon, such as the Magnetic Reconnection Experiment (MRX) built in 1995 at the Princeton Plasma Physics Laboratory. Reconnection is actually a fairly general term that is also used in fluid mechanics when the topology of the vorticity lines is modified. Nowadays, the reconnection process is even observed between quantized vorticies in superfluid helium (Bewley *et al.*, 2008).

7.1 A Current Sheet in Ideal MHD

We will first consider the two-dimensional stationary magnetic configuration of Figure 7.2 where magnetic field lines of opposite directions are separated by a distance 2ℓ. For example, we can think about two close solar magnetic loops (see Figure 3.4). This external configuration being imposed, one wants to know the properties of the inner region of thickness 2ℓ. We will assume that the norm of

Figure 7.1 Streams of charged particles blasted from the Sun collide with Saturn's magnetic field, creating an aurora on the planet's south pole. Unlike Earth's relatively short-lived auroras, Saturn's auroras can last for days. It is thought that magnetic reconnection plays an important role in the formation of auroras. This image is a combination of ultraviolet and visible-light observations made with the Hubble Space Telescope (HST) on January 28, 2004. Credit NASA/ESA J. Clarke (Boston University) and Z. Levay (STScI).

the magnetic field is constant in the external region. From Maxwell's equations, we have

$$\mu_0 \mathbf{j} = \nabla \times \mathbf{B} = -\frac{\partial B_x}{\partial y}\mathbf{e_z}, \tag{7.1}$$

hence $j_z = -B_0/(\ell\mu_0)$ for $y \in [-\ell, +\ell]$ (we assume that the magnetic field varies linearly in the inner region; see Figure 7.2). In other words, a current sheet of thickness 2ℓ appears between the two regions of different magnetic polarity. This current is even more intense given that the sheet is thin and therefore the regions of different magnetic polarity are close. This observation leads us to conclude that, if there is no mechanism to break this configuration, it is possible in principle to increase the intensity of the current sheet as much as we want by bringing the two regions of different magnetic polarity closer together.

Arising from the presence of electrical currents oriented along the z direction, one has a Laplace force along the y direction. This force acts on the magnetic field

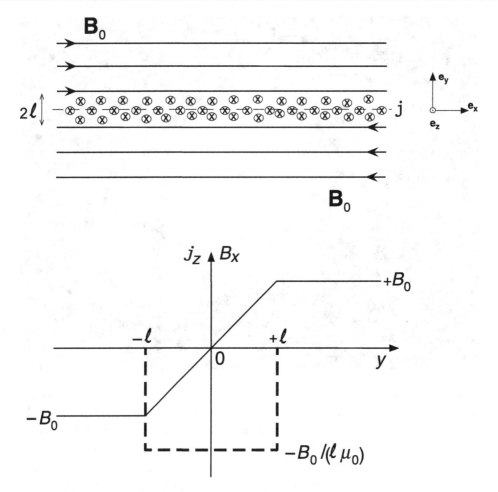

Figure 7.2 Top: current sheet of thickness 2ℓ at the interface of two regions of different magnetic polarity. The configuration is assumed to be invariant along the z direction. Bottom: variations of the magnetic field (solid line) and electric current (dashed line) in the y direction.

lines of the internal region, causing them to converge towards the Ox axis. This force is, however, balanced by the pressure force since, from Eq. (2.20), we have the relationship of static equilibrium

$$\nabla P = \mathbf{j} \times \mathbf{B} = -s(y)\frac{B_0^2}{\ell\mu_0}\mathbf{e_y} , \tag{7.2}$$

with the sign function $s(y) = +1$ for $y > 0$ and $s(y) = -1$ for $y < 0$. Thus, the equilibrium pressure is

$$P(y) = \begin{cases} P_0 - |y|B_0^2/(\ell\mu_0) , & \text{if } y \in [-\ell, +\ell], \\ P_0 - B_0^2/\mu_0 , & \text{otherwise.} \end{cases} \tag{7.3}$$

Thus, the pressure is maximum in the neutral layer, i.e. in the xOz plane. Moreover, Maxwell's equations give us in the stationary regime

$$\nabla \cdot \mathbf{E} = 0 \quad \text{and} \quad \nabla \times \mathbf{E} = \mathbf{0}, \tag{7.4}$$

which implies that the electric field is constant throughout the space. We now explicitly use the hypothesis of zero magnetic diffusion, $\eta = 0$. From the ideal form of Ohm's law, we can conclude that the electric field is actually null throughout the space.

In a collisional plasma where the magnetic diffusion is not zero, the energy will dissipate. We know from relation (3.22) that the magnetic dissipation is proportional to the square of the electric current, therefore one must expect a significant dissipation in the inner region and no dissipation outside. The simple image of the current sheet of Figure 7.2 is thus no longer applicable in the presence of dissipation: Alfvén's theorem is violated and the plasma can move through the magnetic field lines, which can change topology by a magnetic reconnection process.

7.2 The Sweet–Parker Model

The first magnetic reconnection model was proposed by Sweet and Parker in the years 1956–1957 (Parker, 1957; Sweet, 1958). This two-dimensional model, based on the incompressible standard MHD, assumes that the current sheet introduced in the previous section is similar to a diffusion region in which the magnetic reconnection occurs. This region is assumed to be rectangular, with a size of $2L \times 2\ell$ (see Figure 7.3). It is supposed that the configuration is stationary and that the diffusion region is constantly fed by a magnetic field converging at the speed $\mathbf{u}_1$ which, after reconnection, is ejected laterally at the speed $\mathbf{u}_2$. Originally,

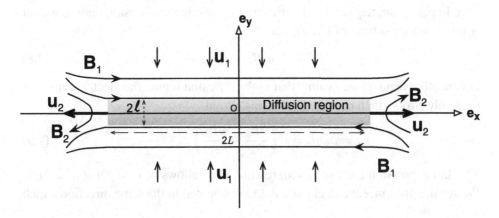

Figure 7.3 Two-dimensional Sweet–Parker model of magnetic reconnection. The reconnection process takes place in the diffusion area (dark region).

the main objective of this model was to explain the explosive nature of solar flares whose characteristic time is of the order of 10 s. Indeed, if we consider only a stationary magnetic diffusion process, the induction equation tells us that the rate of diffusion in the solar corona is

$$u_{\text{diff}} \simeq \frac{\eta}{L} = 10^{-7} \text{ m s}^{-1}, \tag{7.5}$$

where we took $\eta = 1 \text{ m}^2 \text{ s}^{-1}$ and $L = 10^7$ m for a magnetic loop. The characteristic diffusion time is therefore $\tau_{\text{diff}} = L/u_{\text{diff}} \simeq 10^{14}$ s, which is roughly a million years! It is therefore necessary to find another much more efficient mechanism to account for the observations. Before presenting the Sweet–Parker model, it is interesting to evaluate the second characteristic speed of this configuration. Since the medium is assumed to be incompressible, the second natural speed that emerges from the problem is the Alfvén speed: in the solar corona, we have $B = 10^{-2}$ T and $\rho_0 = 10^{-12} \text{ kg m}^{-3}$, hence the Alfvén speed $b \simeq 10^7 \text{ m s}^{-1}$. The information propagated by an Alfvén wave on a length $L = 10^7$ m is therefore done in a time $\tau_A \simeq 1$ s. The objective of our study is to find a reconnection mechanism whose characteristic time τ_{rec} will be such that

$$\tau_A \leq \tau_{\text{rec}} \ll \tau_{\text{diff}}. \tag{7.6}$$

In passing, we see that the **Lundquist number**,[1] $S \equiv Lb/\eta$, trivially connects the Alfvén and diffusion speeds, $b = Su_{\text{diff}}$, and in a similar way the associated time scales,

$$\tau_{\text{diff}} = S\tau_A. \tag{7.7}$$

This last relationship illustrates the large time-scale difference that exists in a turbulent environment ($S \sim 10^{12}$–10^{14} in the solar corona).

The configuration of Figure 7.3 being assumed to be stationary, Maxwell's equations tell us that the electric field is constant throughtout the space in question. In particular, neglecting electric current outside the diffusion region, we get from the standard form of Ohm's law

$$E_z = -u_1 B_1 = -u_2 B_2. \tag{7.8}$$

On the other hand, if we assume that in the diffusion region the electric current is relatively strong, then one obtains for this region

$$E_z = \mu_0 \eta j_z = \eta \left(\frac{\partial B_y}{\partial x} - \frac{\partial B_x}{\partial y} \right) \simeq -\eta \frac{B_1}{\ell}. \tag{7.9}$$

This last expression combined with relation (7.8) allows us to say that $u_1 \simeq \eta/\ell$. We see that the current and electric field are oriented in the same direction which

---

[1] We should not confuse the Lundquist number with the magnetic Reynolds number whose definition is $R_m = Lu/\eta$.

means, according to Eqs. (7.17) and (7.18), that the magnetic energy is transformed into kinetic energy via the electromagnetic power density $\mathbf{j} \cdot \mathbf{E}$. We see also that the direction of the electric current combined with the output magnetic field must give a Laplace force that will eject the plasma from the diffusion region.

The local condition of incompressibility integrated over the diffusion region gives us the relationship (after applying Ostrogradsky's theorem in two dimensions)

$$u_1 L = u_2 \ell . \tag{7.10}$$

The combination of relations (7.8) and (7.10) gives finally $B_1 \ell = B_2 L$. For further analysis, we use the (stationary) equation of motion projected on the Ox axis at $y = 0$, and then on the Oy axis at $x = 0$. One obtains, respectively,

$$\frac{1}{2} \frac{\partial \rho_0 u_x^2}{\partial x} = -\frac{\partial P}{\partial x} - \frac{1}{2\mu_0} \frac{\partial B_y^2}{\partial x} , \tag{7.11}$$

$$\frac{1}{2} \frac{\partial \rho_0 u_y^2}{\partial y} = -\frac{\partial P}{\partial y} - \frac{1}{2\mu_0} \frac{\partial B_x^2}{\partial y} , \tag{7.12}$$

which gives, after integration,

$$\frac{\rho_0 u_x^2}{2} + P + \frac{B_y^2}{2\mu_0} = P(0,0) , \tag{7.13}$$

$$\frac{\rho_0 u_y^2}{2} + P + \frac{B_x^2}{2\mu_0} = P(0,0) , \tag{7.14}$$

with $P(0,0)$ the pressure at the null point – also called the X-point – i.e. the point of convergence of the magnetic field lines (for us, the point of origin) where the velocity and the magnetic field vanish. In particular, we get

$$\frac{\rho_0 u_1^2}{2} + P_1 + \frac{B_1^2}{2\mu_0} = \frac{\rho_0 u_2^2}{2} + P_2 + \frac{B_2^2}{2\mu_0} , \tag{7.15}$$

which gives after simplification (the pressure is assumed to be constant[2] in the external region and $\ell \ll L$)

$$u_2 = b_1 \equiv \frac{B_1}{\sqrt{\rho_0 \mu_0}} . \tag{7.16}$$

Therefore, the plasma is ejected after the magnetic reconnection at the injection Alfvén velocity. We may conclude that the **reconnection rate** – which is by definition u_1/b_1 – is

$$\frac{u_1}{b_1} = \frac{u_1}{u_2} = \frac{\ell}{L} = \sqrt{\frac{\ell}{L} \frac{u_1}{u_2}} = \sqrt{\frac{\ell}{L} \frac{\eta}{\ell b_1}} = S^{-1/2} . \tag{7.17}$$

---

[2] In a highly magnetized region, we can just make the assumption that the pressure is dominated by the magnetic pressure.

One finds, of course, the idea that the more turbulent the system, the thinner the diffusion region. We can deduce that the Sweet–Parker reconnection time is

$$\tau_{\text{rec}} = \frac{L}{u_1} = \tau_{\text{A}} \mathcal{S}^{1/2}. \tag{7.18}$$

Insofar as the Lundquist number is of the order of 10^{14}, we see that this reconnection process is definitely too slow to account for the observations on solar flares. But this time is still much shorter than the diffusion time since, according to expression (7.7), we have

$$\tau_{\text{rec}} = \tau_{\text{diff}} \mathcal{S}^{-1/2}. \tag{7.19}$$

The Sweet–Parker prediction has been reproduced numerically, and experimentally in 1999 at the MRX in Princeton (see Figure 7.4), where the Lundquist number was $\mathcal{S} \sim 1000$.

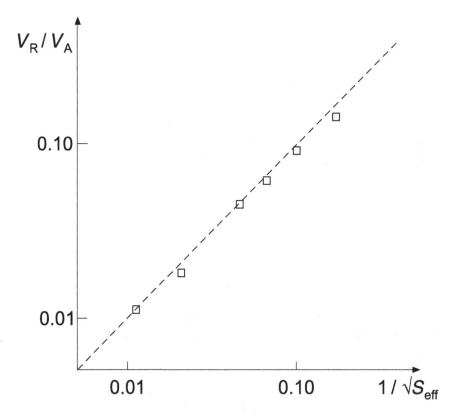

Figure 7.4 Experimental reproduction of the (modified) Sweet–Parker law on the MRX at Princeton. The measured velocities V_R and V_A correspond, respectively, to u_1 and b_1. Adapted from Ji *et al.* (1999).

7.3   Collisionless Hall MHD Reconnection

In order to improve the magnetic reconnection model proposed by Sweet and Parker which has too slow a rate of magnetic reconnection, a new model was proposed in 1964 by Petschek. This model, which is based on the presence of shocks, predicts a fast rate of magnetic reconnection. Numerical simulations (Biskamp, 1986, 2000) demonstrate, however, that this configuration remains stable only if the magnetic diffusion increases near the X-point (this is called anomalous resistivity). An inhomogeneity of magnetic diffusion seems difficult to justify, so this model has been more or less abandoned.

Then, it was natural to study the consequences of the decoupling between ions and electrons for the magnetic reconnection. In this context Hall MHD is often used in a first approximation (Bhattacharjee, 2004). The term **collisionless reconnection** is generally used for this description because Alfvén's theorem is violated not because of the magnetic diffusion – and hence collisions – but because of the bi-fluid character of the plasma in which ions are no longer frozen in the magnetic field. The reconnection configuration gives rise to several typical

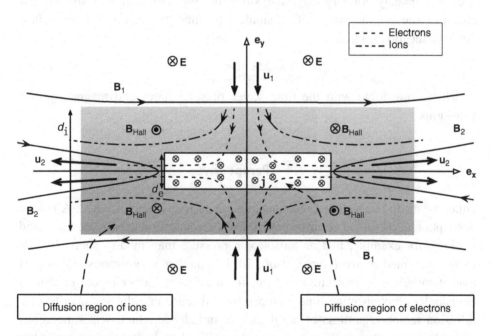

Figure 7.5 The configuration of collisionless magnetic reconnection. In the diffusion region (shaded part) of thickness $\sim d_i$ (the ion inertial length) ions and electrons decouple. The electric current, mainly carried by electrons, creates a quadrupole induced magnetic field $\mathbf{B}_{Hall}$. The reconnection takes place in the diffusion region of electrons (unshaded part) of thickness $\sim d_e$ (the electron inertial length). For example, the thickness of the electron diffusion region is about 1 to 10 km in the magnetosphere and about 1 m in the solar corona.

length scales as seen in Figure 7.5: there is the ion inertial length d_i within which ions unfreeze from the magnetic field and the electron inertial length d_e within which the reconnection actually occurs with the diffusion of electrons. Simplifying, we can say that the electric current is mainly carried by electrons in the diffusion region of ions, which leads to the generation of an induced quadrupole magnetic field (the Hall effect)[3]. As in the Sweet–Parker model, after the magnetic reconnection the plasma is ejected laterally. The main feature of collisionless reconnection is this quadrupole magnetic field aspect. The numerical simulations (Mandt *et al.*, 1994), and the magnetospheric (Mozer *et al.*, 2002) and laboratory (Yamada *et al.*, 2006) measurements with the MRX at Princeton highlight this new property.

The estimates made for the reconnection rate show that it seems to be independent of the magnetic diffusion, with $u_1 \simeq (d_i/L)b_1$, where b_1 is the Alfvén speed at the entrance of the diffusion zone of ions and L is the length of this area (of the order of 10 times d_i). Unlike in the Sweet–Parker model, where the diffusion region becomes increasingly thin when the level of turbulence increases, this thinning finds a finite limit in Hall MHD that is related to the ion inertial length. This is ultimately not very surprising since the Hall term brings to MHD a new characteristic length scale, d_i. Collisionless reconnection can therefore produce fast reconnection such as

$$\boxed{\tau_{\text{rec}} \simeq 10\,\tau_A\,,} \tag{7.20}$$

which is compatible with the time scales of solar flares and magnetospheric substorms.

7.4　Perspectives

Other means to produce fast reconnection are currently being studied. A kinetic description allows us to describe collisionless reconnection even more finely and to show, for example, that an anisotropic pressure may modify somewhat the image presented above (Aunai *et al.*, 2011). Another very promising study is that of turbulent reconnection. Indeed, for the Sweet–Parker model or that of collisionless reconnection, one assumes the existence and the stationarity of a diffusion region regardless of the degree of turbulence. Numerical simulations show, however, that such a region is destabilized at high Lundquist number ($S \geq 1000$) (Daughton *et al.*, 2009). In the turbulent regime we clearly see (Figure 7.6) the appearance of a myriad of current sheets between magnetic eddies

---

[3] To be more precise and to complete Figure 7.5, it should be added that electrons arrive laterally along the magnetic field lines to compensate for the pressure drop associated with the escape of ions. The associated current contributes to the generation of the induced quadrupole magnetic field.

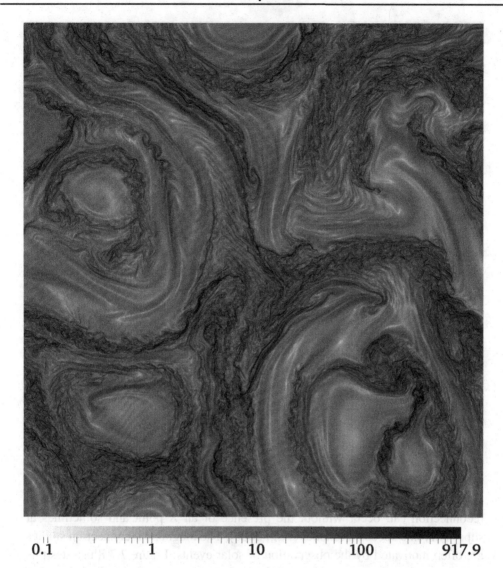

0.1   1   10   100   917.9

Figure 7.6 Direct numerical simulation of MHD turbulence at a spatial resolution of $2048 \times 2048 \times 128$ points. This is a snapshot of the current density modulus (on a logarithmic scale) in the presence of a relatively strong uniform magnetic field in the direction transverse to the picture (and for which the resolution is lower). Only a fraction of the plane simulated is shown. Current sheets are localized at the interface of magnetic eddies where magnetic reconnection may happen. Courtesy of Dr. R. Meyrand.

where magnetic reconnection may happen. Furthermore, recent statistical studies suggest that MHD reconnection tends also to become fast and independent of the magnetic diffusion when the regime of fully developed turbulence is reached (Servidio *et al.*, 2009). The small-scale nature of the magnetic reconnection is still poorly understood. For that reason, the Magnetospheric Multiscale Mission

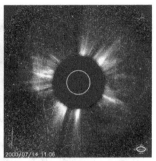

Figure 7.7 Images (in the visible domain) of the Sun taken by the coronagraph LASCO/SoHO during the solar flare called "Bastille Day" (July 14, 1998). The inner circle is the size of the Sun. The image shows a coronal mass ejection produced during this flare, which is considered to be one of the largest ever observed. Credit ESA/NASA. The detailed study of this event shows the presence of an X-point (Aulanier *et al.*, 2000).

(MMS/NASA) – launched in 2015 – has been designed to gather information about the microphysics at electron scales of the magnetic reconnection near Earth. The four identically instrumented spacecraft fly in an adjustable pyramid-like formation that enables them to observe the three-dimensional structure of magnetic reconnection, with unprecedented time resolution and the accuracy needed to capture the elusive thin and fast-moving electron diffusion region.

To conclude, it is important to realize that the two-dimensional configuration presented in this chapter is *a priori* relevant because a plasma spontaneously generates current sheets. However, the three-dimensional nature of the problem may give access to important additional physical properties: with the help of numerical simulations one can show, for example, that a slip-running magnetic reconnection can occur without the presence of an X-point and sometimes at super-Alfvénic speeds (Aulanier *et al.*, 2006). The magnetic reconnection studies are often motivated by the observation of solar events: Figure 7.7 illustrates this with the large-scale consequences of a solar flare that was probably triggered after a magnetic reconnection process.

Exercises for Part II

II.a Modes g and p

We consider a medium subject to the gravitational force, $-\rho \nabla \Psi$, such that

$$\psi = 4\pi \mathcal{G}\rho$$

is the gravitational potential and $\mathcal{G}$ is the gravitational constant. Find the modes g and p associated with a perturbation of the medium.

II.b Magnetostrophic Waves

In the case of the geodynamo problem we know that rotation plays a central role in the regeneration of the large-scale magnetic field because the Rossby number is very small ($\sim 10^{-6}$). Since liquid metals are weakly compressible, we can consider the incompressible MHD system. Upon introducing the Coriolis force into the incompressible MHD equations, one obtains (a uniform magnetic field $\mathbf{b}_0$ is also included to mimic the dipolar field)

$$\frac{\partial \mathbf{u}}{\partial t} + 2\Omega_0 \times \mathbf{u} + \mathbf{u} \cdot \nabla \mathbf{u} = -\nabla P_* + \mathbf{b}_0 \cdot \nabla \mathbf{b} + \mathbf{b} \cdot \nabla \mathbf{b} + \nu \, \Delta \mathbf{u},$$

$$\frac{\partial \mathbf{b}}{\partial t} + \mathbf{u} \cdot \nabla \mathbf{b} = \mathbf{b}_0 \cdot \nabla \mathbf{u} + \mathbf{b} \cdot \nabla \mathbf{u} + \eta \, \Delta \mathbf{b},$$

where Ω_0 is the rotation rate, which is assumed to be uniform and parallel to $\mathbf{b}_0$.

(1) Linearize the (inviscid and ideal) system and find the dispersion relation.
(2) Plot the dispersion relation with a suitable normalization.
(3) Specify the physical meaning of the magnetostrophic waves.

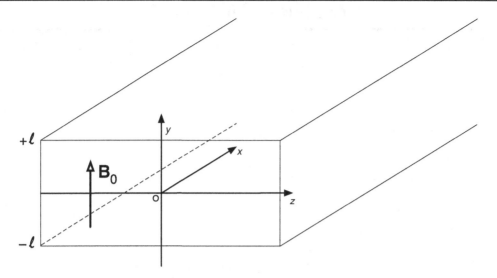

Figure II.c.1 Liquid metal flowing between the insulating plates forming an infinitely long parallelepiped in the Ox direction.

II.c Liquid Metals and the Hartmann layer

The motion of a liquid metal under a strong magnetic field is a subject of important studies in the domain of industrial MHD. Indeed, the magnetic field is often used to heat or mix a liquid metal. In this problem, we will consider a stationary incompressible flow between insulating plates forming an infinitely long parallelepiped in the Ox direction (see Figure II.c.1). We will use the standard incompressible MHD and assume that a constant pressure force applies in the Ox direction to maintain a steady flow. The width of the parallelepiped is large enough for us to forget the z dependence: the problem is thus reduced to the single variable y. A uniform external magnetic field $\mathbf{b}_0$ is imposed in the y direction.

(1) Show that the presence of insulating plates imposes the condition $b_x(\pm\ell) = 0$.

(2) Demonstrate that the MHD equations reduce to the following system:

$$\frac{d^2 u_x(y)}{dy^2} + \frac{b_0}{\nu}\frac{db_x(y)}{dy} = C,$$

$$\frac{d^2 b_x(y)}{dy^2} + \frac{b_0}{\eta}\frac{du_x(y)}{dy} = 0,$$

where C is a constant that we will determine.

(3) Find the velocity profile solution of the problem. We will introduce in the expression the Hartmann number:

$$H_a \equiv \frac{b_0 \ell}{\sqrt{\nu \eta}} .$$

(4) Discuss the velocity profile according to the value of the Hartmann number. In which case can we find the Poiseuille parabolic profile?

II.d Magnetic Field Lines and the X-Point

Sketch the magnetic field lines (with their orientations) in two dimensions for $\mathbf{B} = (y, a^2 x)$, where a is a parameter. Give the magnetic pressure force and the tension force. In which case is the current density null?

Part III

Instabilities and Magnetic Confinement

8

Static Equilibrium

The control of the magnetic confinement of a plasma in a tokamak is a major challenge for the success of thermonuclear fusion. Indeed, the confinement allows one to maintain a plasma at high temperature while avoiding leakage of particles by diffusion, which, moreover, can damage the walls of the reactor. Within this framework, the first step of the analysis consists of the determination of the plasma equilibrium state from which a stability analysis can be carried out. In this part, we shall consider the standard, ideal, and inviscid MHD equations. This chapter will be devoted to the static equilibrium for different geometries, then Chapter 9 will present the linear perturbation theory, allowing us finally to solve a few problems of stability in Chapter 10.

8.1 Equilibrium Equations

The standard, ideal, and inviscid MHD equations in the case of a static equilibrium can be written as

$$\mathbf{u} = 0, \tag{8.1}$$

$$0 = -\nabla P + \frac{1}{\mu_0}(\nabla \times \mathbf{B}) \times \mathbf{B}, \tag{8.2}$$

$$\nabla \cdot \mathbf{B} = 0. \tag{8.3}$$

The plasma equilibrium is defined by the condition $\partial/\partial t = 0$. If the equilibrium is static the plasma does not move and we also have the condition (8.1). The second relationship (8.2) comes from the simplification of the equation of motion. Using classical notation for this type of problem, we define the state of static equilibrium by the fields $P_0(\mathbf{r})$, $\mathbf{j}_0(\mathbf{r})$, and $\mathbf{B}_0(\mathbf{r})$ that satisfy the equations

$$\nabla P_0 = \mathbf{j}_0 \times \mathbf{B}_0, \tag{8.4}$$

$$\nabla \cdot \mathbf{B}_0 = 0, \tag{8.5}$$

$$\mathbf{j}_0 = \frac{1}{\mu_0}(\nabla \times \mathbf{B}_0). \tag{8.6}$$

The fields considered are therefore independent of time, but they are not necessarily constant since they may depend on space. We see immediately that

$$\mathbf{B}_0 \cdot \nabla P_0 = 0 \,, \tag{8.7}$$

$$\mathbf{j}_0 \cdot \nabla P_0 = 0 \,, \tag{8.8}$$

which means that the magnetic field lines and the electric current lines are localized on surfaces perpendicular to the pressure gradient. In other words, the surfaces at constant pressure (isobars) are any magnetic or current surfaces, i.e. surfaces that are tangent to the magnetic field or to the current. Note that this does not mean that the magnetic field and the electric current are necessarily parallel on the isobars.

8.2   Magnetic Confinement by θ-Pinch

We will consider the static equilibrium of a magnetic confinement by a θ-pinch, i.e. the situation where the plasma is confined by an intense azimuthal electric current whose effect is primarily to generate an axial magnetic field (see Figure 8.1). It is the direction of the current of confinement which gave the name

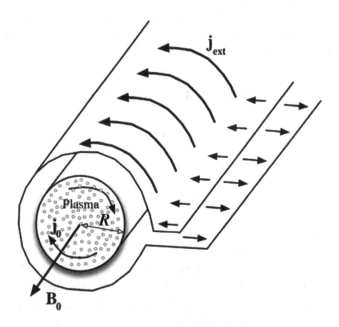

Figure 8.1 A schematic view of magnetic confinement by a θ-pinch. An intense azimuthal electric current $\mathbf{j}_{ext}$ is generated in the outer cylindrical envelope. As a result, the magnetic field $\mathbf{B}_0$ which crosses the plasma (spheres) in the cylinder of radius R is predominantly axial. The electric current in the plasma is also azimuthal but opposite to the external current such that the Laplace force confines the plasma.

to this type of machine (θ-pinch configuration). The geometric configuration has a cylindrical symmetry, so the variables are independent of θ and z (the length of the cylinder). Under these conditions, we have the following relation (in cylindrical coordinates) between the electric current and the induced magnetic field:

$$\mathbf{j}_0 = \frac{1}{\mu_0}\left[0, -\frac{\partial B_{0z}}{\partial r}, \frac{1}{r}\frac{\partial(rB_{0\theta})}{\partial r}\right]. \tag{8.9}$$

Then, the static equilibrium can be written as

$$\nabla P_0 = \frac{1}{\mu_0}\left[-B_{0z}\frac{\partial B_{0z}}{\partial r} - \frac{B_{0\theta}}{r}\frac{\partial(rB_{0\theta})}{\partial r}, \frac{B_{0r}}{r}\frac{\partial(rB_{0\theta})}{\partial r}, B_{0r}\frac{\partial B_{0z}}{\partial r}\right]. \tag{8.10}$$

Insofar as the magnetic field of confinement is supposed to be relatively strong, the axial component will give the main contribution (this is a similar situation to that of a solenoid), hence the plasma equilibrium equation at main order is

$$\frac{\partial P_0}{\partial r} = -\frac{1}{\mu_0}B_{0z}\frac{\partial B_{0z}}{\partial r}. \tag{8.11}$$

We can check in passing that the Laplace force efficiently confines the plasma because of its orientation along the direction of the cylinder axis. The equilibrium equation may be rewritten as

$$\frac{\partial}{\partial r}\left(P_0 + \frac{B_{0z}^2}{2\mu_0}\right) = 0; \tag{8.12}$$

its solution is given by the boundary condition (zero pressure on the external edge)

$$P_0(r) + \frac{B_{0z}^2(r)}{2\mu_0} = \frac{B_{0z}^2(R)}{2\mu_0}. \tag{8.13}$$

We see that only the magnetic pressure (not the magnetic tension) is involved in the equilibrium. In introducing the pressure ratio, $\beta(r) \equiv 2\mu_0 P_0(r)/B_{0z}^2(R)$, we finally get the solution

$$\boxed{B_{0z}(r) = B_{0z}(R)\sqrt{1 - \beta(r)}.} \tag{8.14}$$

For thermonuclear plasmas we have generally $\beta \sim 10^{-2}$–10^{-1}, with a decreasing kinetic pressure on approaching the edge of the cylinder. We will see later (Chapter 10) that this configuration is stable; however, its use is limited because particles can escape from both ends of the cylinder at the thermal speed.

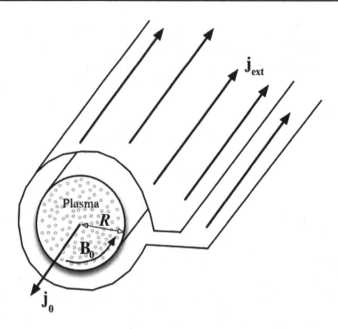

Figure 8.2 A schematic view of magnetic confinement by a z-pinch. An intense axial electric current $\mathbf{j}_{\text{ext}}$ is generated in the external cylindrical envelope. As a result, the magnetic field $\mathbf{B}_0$ which crosses the plasma (spheres) in the cylinder of radius R is mainly azimuthal. The electric current in the plasma is also axial, in the direction opposite to the external current such that the Laplace force confines the plasma.

8.3   Magnetic Confinement by z-Pinch

Now consider the static equilibrium of magnetic confinement by a z-pinch, i.e. the situation where the plasma is confined by a strong axial electric current whose effect is to generate primarily an azimuthal magnetic field (see Figure 8.2). The geometric configuration is still of cylindrical symmetry, which gives at main order

$$\nabla P_0 = \frac{1}{\mu_0}\left[-\frac{B_{0\theta}}{r}\frac{\partial(rB_{0\theta})}{\partial r}, 0, 0\right]. \tag{8.15}$$

The Laplace force is therefore oriented towards the cylinder axis if the derivative is positive. The equilibrium equation can be written as

$$\frac{\partial}{\partial r}\left[P_0 + \frac{B_{0\theta}^2}{2\mu_0}\right] = -\frac{B_{0\theta}^2}{\mu_0 r}. \tag{8.16}$$

We see that the magnetic pressure and the magnetic tension are involved in the equilibrium. We will see later (Chapter 10) that this configuration is highly unstable.

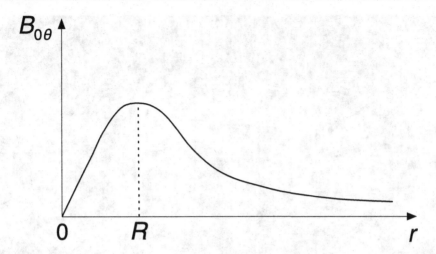

Figure 8.3 Schematic variation of the azimuthal component of the magnetic field for the Bennett pinch.

To illustrate this equilibrium, we will consider the Bennett pinch for which the $B_{0\theta}$ profile is (see Figure 8.3)

$$B_{0\theta}(r) = \frac{\mu_0 I}{2\pi} \frac{r}{r^2 + R^2},$$
(8.17)

where I is the total current which circulates through the pinch. In this case, we get

$$j_{0_z}(r) = \frac{I}{\pi} \frac{R^2}{(r^2 + R^2)^2},$$
(8.18)

which leads to a confinement Laplace force. From the equilibrium Eq. (8.16), we find the solution

$$P_0(r) = \frac{\mu_0 I^2}{8\pi^2} \frac{R^2}{(r^2 + R^2)^2}.$$
(8.19)

Interestingly, it is with such an experiment that the highest ionic temperatures (more than 10^9 K) have been reached recently in Albuquerque (New Mexico) (Haines *et al.*, 2006), which made this place momentarily the hottest point of our galaxy!

8.4   Toroidal Tokamak Configuration

The combination of the θ- and z-pinches leads to a helical magnetic field: this is called screw pinch or simply the right tokamak configuration since it is this type of magnetic field that we find in tokamaks (see Figure 8.4). We have seen that it was not an efficient solution to confine the plasma because of the leakages of

Figure 8.4 Internal view of the Joint European Torus (JET) tokamak built in 1983 near Oxford. Courtesy of EUROfusion.

particles at the edges of the cylinder. The simple solution is to close the cylinder by wrapping it: we get a torus, which is the typical tokamak configuration.

8.4.1 The Grad–Shafranov Equation

To describe the configuration of a toroidal plasma equilibrium, we will use the cylindrical coordinates and assume that the problem is axisymmetric, i.e. there is no dependence on θ. By introducing the **stream function** $\psi \equiv rA_\theta$ (see also Chapter 5), where $\mathbf{A}$ is the vector potential, we get

$$B_r = -\frac{1}{r}\frac{\partial \psi}{\partial z}, \tag{8.20}$$

$$B_z = \frac{1}{r}\frac{\partial \psi}{\partial r}. \tag{8.21}$$

It may be noted that this implies the relationship $\mathbf{B} \cdot \nabla \psi = 0$: in other words, the magnetic surfaces are surfaces at constant ψ. The stream function is in fact related to the poloidal flux through a disk of radius r centered on the origin at the altitude $z = 0$ (see Figure 8.5):

$$\iint \mathbf{B} \cdot d\mathcal{S} = \iint \nabla \times \mathbf{A} \cdot d\mathcal{S} = \oint \mathbf{A} \cdot d\boldsymbol{\ell} = 2\pi rA_\theta = 2\pi \psi(z = 0). \tag{8.22}$$

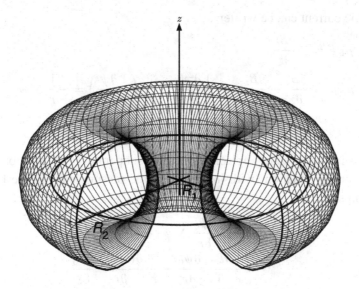

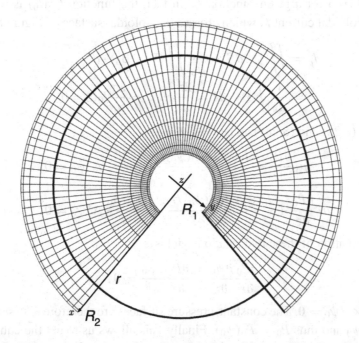

Figure 8.5 Geometry of a torus of small radius R_1 and large radius R_2 centered on O. The poloidal flux is calculated through the disk centered on O, at the altitude $z = 0$ and of radius r such that $R_1 < r < R_2$. In the case of ITER, the expected dimensions are $R_1 = 4.2$ m and $R_2 = 8.2$ m.

The electric current can be written as

$$\mu_0 j_r = -\frac{\partial B_\theta}{\partial z},$$
(8.23)

$$\mu_0 j_\theta = \frac{\partial B_r}{\partial z} - \frac{\partial B_z}{\partial r} = -\frac{1}{r}\frac{\partial^2 \psi}{\partial z^2} - \frac{\partial}{\partial r}\left(\frac{1}{r}\frac{\partial \psi}{\partial r}\right) = -\frac{1}{r}\Delta^*\psi,$$
(8.24)

$$\mu_0 j_z = \frac{1}{r}\frac{\partial(rB_\theta)}{\partial r},$$
(8.25)

where Δ^* is the pseudo-Laplacian or Stokes operator (see relation (5.13)). The equilibrium equation (with indices $_0$) of the plasma leads to the following relationships:

$$\mu_0\frac{\partial P_0}{\partial r} = -\frac{1}{r^2}\frac{\partial \psi_0}{\partial r}\Delta^*\psi_0 - \frac{B_{0\theta}}{r}\frac{\partial(rB_{0\theta})}{\partial r},$$
(8.26)

$$0 = \frac{1}{r}\frac{\partial B_{0\theta}}{\partial z}\frac{\partial \psi_0}{\partial r} - \frac{1}{r^2}\frac{\partial(rB_{0\theta})}{\partial r}\frac{\partial \psi_0}{\partial z},$$
(8.27)

$$\mu_0\frac{\partial P_0}{\partial z} = -\frac{1}{r^2}\frac{\partial \psi_0}{\partial z}\Delta^*\psi_0 - B_{0\theta}\frac{\partial B_{0\theta}}{\partial z}.$$
(8.28)

Relation (8.27) can be rewritten as $\nabla\psi_0 \times \nabla(rB_{0\theta}) = \mathbf{0}$, which means that the surfaces at constant ψ_0 are also surfaces at constant $rB_{0\theta}$, hence the relation $rB_{0\theta} = I_0(\psi_0)$ for a given function I_0. In fact, the function $I_0(\psi_0)$ is related to the total poloidal current I_p which crosses the poloidal surface of Figure 8.5:

$$I_p = \iint \mathbf{j}_0 \cdot d\mathcal{S} = \frac{1}{\mu_0}\iint (\nabla \times \mathbf{B}_0) \cdot d\mathcal{S}$$

$$= \frac{1}{\mu_0}\oint \mathbf{B}_0 \cdot d\boldsymbol{\ell} = \frac{2\pi}{\mu_0}rB_{0\theta} = \frac{2\pi}{\mu_0}I_0(z=0).$$
(8.29)

Relations (8.26) and (8.28) can be rewritten as

$$\mu_0\frac{\partial P_0}{\partial r} + \frac{I_0(\psi_0)}{r^2}\frac{\partial I_0(\psi_0)}{\partial r} + \frac{1}{r^2}\frac{\partial \psi_0}{\partial r}\Delta^*\psi_0 = 0,$$
(8.30)

$$\mu_0\frac{\partial P_0}{\partial z} + \frac{I_0(\psi_0)}{r}\frac{\partial(I_0(\psi_0)/r)}{\partial z} + \frac{1}{r^2}\frac{\partial \psi_0}{\partial z}\Delta^*\psi_0 = 0.$$
(8.31)

The combination of these two relations yields

$$\frac{\partial P_0}{\partial r}\frac{\partial \psi_0}{\partial z} - \frac{\partial P_0}{\partial z}\frac{\partial \psi_0}{\partial r} = 0,$$
(8.32)

i.e. $\nabla P_0 \times \nabla\psi_0 = \mathbf{0}$. The constant-pressure surfaces are therefore also surfaces at constant ψ_0 and thus $P_0 = P_0(\psi_0)$. Finally, this allows us to get the equilibrium equation of a toroidal plasma:

$$\boxed{\mu_0 r^2\frac{\partial P_0}{\partial \psi_0} + I_0\frac{\partial I_0}{\partial \psi_0} + \Delta^*\psi_0 = 0.}$$
(8.33)

Relation (8.33) is known as the **Grad–Shafranov equation**. It is a non-linear equation which expresses the balance between the plasma pressure (the first term) and the strength of the Laplace force (the second and third terms). More specifically, the second term can be identified with the poloidal current and toroidal magnetic field (θ-pinch configuration), while the third term corresponds to the toroidal current and poloidal magnetic field (z-pinch configuration). One can verify that, with the elimination of the third term, it is possible to integrate the Grad–Shafranov equation and get the θ-pinch solution (8.13).

8.4.2   The Soloviev Exact Solution

The Grad–Shafranov equation is probably the most well-known MHD equation in the field of fusion plasmas. It was the subject of numerous analytical and especially numerical studies because the solutions are usually not trivial due to the non-linear nature of the equation (Goedbloed *et al.*, 2010). In practice, the nature of the solution depends on the choice of the two arbitrary functions $I_0(\psi_0)$ and $P_0(\psi_0)$, and of the boundary conditions. The resolution of the problem allows one to get the stream function $\psi_0(r, z)$. To get an idea of the type of possible solutions of the Grad–Shafranov equation, we will consider the Soloviev equilibrium, which has the advantage of being analytical. This solution takes the form

$$\psi_0(r, z) = Az^2(r^2 + BR_m^2) + C(r^2 - R_m^2)^2 . \tag{8.34}$$

The introduction of this solution into (8.35) gives

$$\mu_0 r^2 \frac{\partial P_0}{\partial \psi_0} + I_0 \frac{\partial I_0}{\partial \psi_0} + (8C + 2A)r^2 + 2ABR_m^2 = 0 . \tag{8.35}$$

Therefore, we see that the Soloviev solution can be obtained by imposing the conditions:

$$P_0(\psi_0) = D - \frac{1}{\mu_0}(8C + 2A)\psi_0 , \tag{8.36}$$

$$I_0^2(\psi_0) = E - 4ABR_m^2 \psi_0 , \tag{8.37}$$

where A, B, C, D, and E are constants. These conditions thus correspond to linear profiles for $P_0(\psi_0)$ and $I_0^2(\psi_0)$. We can also notice that the solution (8.34) gives a zero flux on the magnetic axis ($r = R_m$, $z = 0$): this must be seen as a boundary condition that we have imposed. It is easy to find the shape of local solutions around the magnetic axis. To do this, let us define

$$r = R_m + \epsilon r , \tag{8.38}$$

$$z = \epsilon z , \tag{8.39}$$

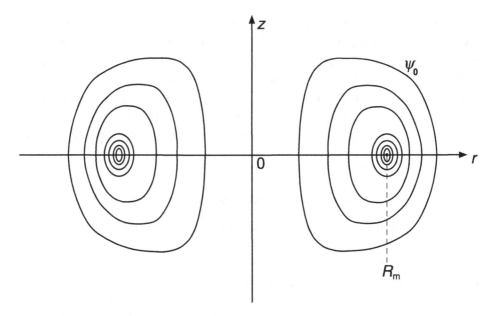

Figure 8.6 Schematic representation of flux isocontours ψ_0 for the Soloviev equilibrium solution in a tokamak. The magnetic axis for which $\psi_0 = 0$ corresponds to $r = R_m$.

with $\epsilon \ll 1$. One obtains at the main order

$$\psi_0(r, z) = \epsilon^2 R_m^2 [A(1 + B)z^2 + 4Cr^2]. \tag{8.40}$$

This is the equation of an ellipse centered on the magnetic axis. It can be seen that the isocontours at constant flux correspond necessarily to a small flux ($\psi_0 \sim \varepsilon^2$). In the general case the form of the isocontours is non-trivial. If one seeks the point of intersection of the isocontours with the radial axis, one obtains (for $r \geq 0$)

$$r_{\pm} = \sqrt{\left| R_m^2 \pm \sqrt{\left| \frac{\psi_0}{C} \right|} \right|}. \tag{8.41}$$

For example, if $\sqrt{|\psi_0/C|} = R_m^2/2$, then $r_- = R_m/\sqrt{2} \simeq 0.7R_m$ and $r_+ = \sqrt{3/2}R_m \simeq 1.2R_m$: therefore, this isocontour is not centered on the magnetic axis. Figure 8.6 shows schematically some solutions of the Soloviev equilibrium. One sees that the magnetic axis does not match with the poloidal geometric axis: we call this difference the Shafranov shift (this term was originally introduced for circular isocontours).

8.5   Force-Free Fields

We conclude this chapter with a type of equilibrium that is very useful in astrophysics and in particular in solar physics: the force-free fields, also called Beltrami fields, which apply when the plasma β is small (i.e. for a magnetic pressure much higher than P_0). One obtains

$$\frac{1}{\mu_0}(\nabla \times \mathbf{B}_0) \times \mathbf{B}_0 = 0\,, \qquad (8.42)$$

hence the relation

$$\boxed{\nabla \times \mathbf{B}_0 = \alpha(\mathbf{x})\mathbf{B}_0\,,} \qquad (8.43)$$

where α (the inverse of a length) depends in general on the position $\mathbf{x}$. Three cases can be analyzed:

- $\alpha = 0$, potential field for which the electric current is zero;
- $\alpha =$ constant, linear force-free field;
- $\alpha = \alpha(\mathbf{x})$, non-linear force-free field.

The application of the divergence operator to relation (8.43) gives

$$\mathbf{B}_0 \cdot \nabla\alpha(\mathbf{x}) = 0\,, \qquad (8.44)$$

which means that α is constant along the magnetic field lines. Furthermore, the application of the rotational operator to relation (8.43) gives

$$\Delta\mathbf{B}_0 + \alpha^2\mathbf{B}_0 = \mathbf{B}_0 \times \nabla\alpha\,. \qquad (8.45)$$

In particular, we see that, for a linear force-free field, expression (8.45) reduces to the Helmholtz equation, whose solutions may be obtained by classical methods. To these equations we have to add the boundary conditions in order to solve the system. In the most general case (with $\alpha(\mathbf{x})$ and in three dimensions), the problem is not trivial and it is necessary to use computers. There is, however, an important result that we can easily establish. Let us assume a volume of plasma $\mathcal{V}$ in which a magnetic field $\mathbf{B}$ is present whose value is zero at the surface. We decompose this field into the form

$$\mathbf{B}_0 = \mathbf{B}_{\text{pot}} + \mathbf{B}_1\,, \qquad (8.46)$$

where $\mathbf{B}_{\text{pot}} = -\nabla\Phi$ is a potential field (satisfying the condition $\Delta\Phi = 0$). By construction the field $\mathbf{B}_1$ is zero at the surface of the volume. Thus, the magnetic contribution at the surface comes from the potential field only: if this contribution is null, we can easily show that the potential field is null everywhere in the volume. The magnetic energy of the system at the equilibrium can be written as

$$E_0^{\text{m}} = \iiint \frac{(\mathbf{B}_{\text{pot}} + \mathbf{B}_1)^2}{2\mu_0}\,d\mathcal{V}$$

$$= \frac{1}{2\mu_0}\iiint(-\mathbf{B}_{\text{pot}} \cdot \nabla\Phi + \mathbf{B}_1^2 - 2\mathbf{B}_1 \cdot \nabla\Phi)d\mathcal{V}$$

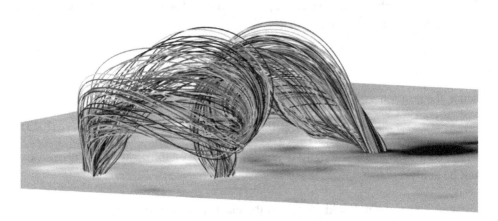

Figure 8.7 Coronal structures re-built under the hypothesis of a non-linear force-free field. The surface gives the direction of the axial photospheric magnetic field measured by the spacecraft Hinode (outward magnetic field in white; inward magnetic field in black). Courtesy of A. Canou. (See also Canou and Amari (2010).)

$$
\begin{aligned}
&= \frac{1}{2\mu_0} \left[-\oiint \Phi \mathbf{B}_{\text{pot}} \cdot d\mathcal{S} + \iiint \mathbf{B}_1^2 \, d\mathcal{V} - 2 \oiint \Phi \mathbf{B}_1 \cdot d\mathcal{S} \right] \\
&= \frac{1}{4\mu_0} \oiint \nabla \Phi^2 \cdot d\mathcal{S} + \frac{1}{2\mu_0} \iiint \mathbf{B}_1^2 \, d\mathcal{V} \\
&= E_{\text{pot}}^{\text{m}} + E_1^{\text{m}}.
\end{aligned}
\tag{8.47}
$$

Therefore, $E_0^{\text{m}} \geq E_{\text{pot}}^{\text{m}}$ and the potential field corresponds to a state of minimal magnetic energy. In solar physics, we define E_1^{m} as the free magnetic energy: this is the maximal energy that can be released during a solar flare. Figure 8.7 shows solar coronal structures re-built under the hypothesis of a non-linear force-free field and from the photospheric magnetic field measurements which are used as boundary conditions.

9

Linear Perturbation Theory

When a static equilibrium has been found (see Chapter 8), the next question that we have to address concerns the stability of this equilibrium. A part of the answer is given by the linear perturbation theory, which consists of analyzing the result of a small (i.e. linear) perturbation of the equilibrium. If the equilibrium is stable, the perturbation will behave as a wave that propagates in the medium; if it is unstable, the perturbation will increase exponentially.

9.1 Instabilities

9.1.1 Classification

In Figure 9.1, we present some unstable and stable situations arising from the example of a sphere placed in an external potential field. In case 1 a sphere is at the bottom of a well of infinite potential. In this position the sphere can only perform oscillations around its equilibrium position. These oscillations, once generated, are damped due to friction until the sphere reaches a static equilibrium position at the bottom of the potential well. This is a situation of **stable** equilibrium. In case 2 a sphere is placed a the top of a potential (a hill). In this case, a small displacement of the sphere is sufficient to move it to much lower potentials: this is an unstable situation that is often associated with a **linear** instability. The third case is that of a **metastable** state where the sphere is placed initially on a locally flat potential (a plateau): a small displacement around the initial position does not change the potential of the sphere. Finally, the last case (case 4) is that of a sphere placed in a hollow. This is an example of **non-linear** instability: the sphere is stable against small perturbations but becomes unstable for larger disturbances.

In plasma physics, the sphere in the previous paragraph corresponds to a particular mode of a wave and the shape of the potential can be a source of free energy. There are many energy sources in space plasmas. For example, the solar wind is a continuous source of energy for the Earth's magnetospheric plasma

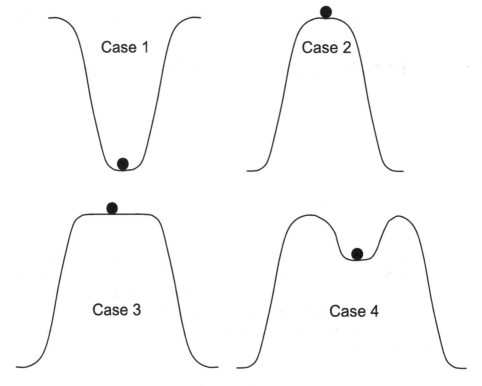

Figure 9.1 Examples of stable and unstable configurations.

which is never in a static equilibrium. The consequences of this energy input are the generation of large-scale gradients and the deformation of the distribution functions of particles at small scales. We speak respectively of macro-instabilities and micro-instabilities. In the case of a tokamak, the stability issue is partly technological: the appearance of instability can lead to de-confinement of the plasma, which, with a temperature of about 10^8 K, can cause severe damage to the walls of the reactor.

This chapter is devoted to the macro-instabilities described by the standard MHD equations. In all cases, we will consider only linear instabilities.

9.1.2 Condition of Existence

The concept of linear instability arises from a formal consideration of the wave function. In the linear wave theory, the amplitude of the waves is much less than the amplitude of the stationary state, so it can be considered as a small and fast perturbation of the medium. The wave function can then be decomposed into Fourier modes such as

$$\delta A(\mathbf{r}, t) = \sum_k A_k \exp(i\mathbf{k} \cdot \mathbf{r} - i\omega t). \tag{9.1}$$

In general, the dispersion relation is a complex equation (in $\mathbb{C}$) whose solution in angular frequency is also complex, with $\omega = \omega_r + i\gamma$ and by definition $\omega_i \equiv \gamma$. From the previous equation, we can clearly see that for a real frequency the perturbation δA of the medium is an oscillating wave. On the other hand, for a complex frequency the amplitude of the wave depends on the sign of the imaginary part γ. If $\gamma < 0$, the real part of the amplitude decreases exponentially with time and the wave is damped. If $\gamma > 0$, the wave amplitude increases exponentially with time: we have a linear instability. In this case γ is called the growth rate. It is important to realize that an instability cannot grow unless there is a free energy, usually a source of energy, available in the plasma. Otherwise the resulting solution is not physically feasible and leads to a violation of energy conservation.

If the growth rate is non-zero, the amplitude of an unstable mode increases as

$$A_k(t) = A_k \exp(\gamma t). \tag{9.2}$$

The linear approximation is therefore not valid indefinitely and becomes false when the amplitude of the wave is comparable to the value of the unperturbed field, i.e. when $A_k(t) \simeq A_0$, or equivalently when the time becomes comparable to the non-linear time:

$$t_{\mathrm{NL}} \simeq \gamma^{-1} \ln\left(\frac{A_0}{A_k}\right). \tag{9.3}$$

The linear approximation of the unstable modes is therefore valid for a time $t \ll t_{\mathrm{NL}}$. When the linear approximation is violated other processes come into play; these processes are non-linear. The time t_{NL} is reached even faster than the growth rate increases. When it becomes larger than the angular frequency ($\gamma > \omega_r$), the wave amplitude explodes and the wave does not have the time to make even a single oscillation; then the concept of a wave becomes obsolete. It is therefore necessary to consider only instabilities whose growth rate is relatively low, i.e. when

$$\gamma/\omega_r \ll 1, \tag{9.4}$$

so that the linear approximation remains valid over several periods of oscillation.

9.2 Kinetic Versus Fluid

9.2.1 The Kinetic Approach

Instability can be strong or weak. Strong instability has a growth rate which violates the inequality (9.4) and coincides, in many cases, with non-oscillatory instabilities. For weak instabilities that satisfy relation (9.4), one can define the general algorithm for obtaining the growth rate from the dispersion relation of the medium. This method is typically used for the study of kinetic instabilities from Vlasov's equation, Eq. (1.10). The dispersion relation, $D(\mathbf{k}, \omega) = 0$, is an implicit relationship between the angular frequency and its wavenumber; given a wavenumber (assumed real) it is possible to determine the angular frequency.

In general, $D(\mathbf{k}, \omega)$ is a complex function such as

$$D(\mathbf{k}, \omega) = D_\mathrm{r}(\mathbf{k}, \omega) + iD_\mathrm{i}(\mathbf{k}, \omega) \,. \tag{9.5}$$

This gives us two equations that can be used to determine ω. In the case of weak instabilities one can make a Taylor expansion, at first order, of $D(\mathbf{k}, \omega)$ (whose complex part is supposed to remain small) around the real part of the frequency ω_r. This gives

$$D(\mathbf{k}, \omega) \simeq D_\mathrm{r}(\mathbf{k}, \omega_\mathrm{r}) + (\omega - \omega_\mathrm{r}) \frac{\partial D_\mathrm{r}(\mathbf{k}, \omega)}{\partial \omega}\bigg|_{\gamma=0} + iD_\mathrm{i}(\mathbf{k}, \omega_\mathrm{r}) = 0 \,. \tag{9.6}$$

Since, by definition, $\omega - \omega_\mathrm{r} = i\gamma$, this equation allows us to obtain a dispersion relation for the real part of the frequency and the linear growth rate when $D_\mathrm{r}(\mathbf{k}, \omega_\mathrm{r}) = 0$:

$$\gamma \equiv \omega_\mathrm{i} = -\frac{D_\mathrm{i}(\mathbf{k}, \omega_\mathrm{r})}{\partial D_\mathrm{r}(\mathbf{k}, \omega)/\partial \omega|_{\gamma=0}} \,. \tag{9.7}$$

This relation is fundamental in the context of the kinetic approach. We will see later how the problem is expressed in the pure fluid case.

Although the concept of instability is mathematically fairly simple to define, it presents a number of difficulties in physics. Indeed, obtaining a positive imaginary part for the angular frequency does not necessarily imply the finding of instability. Instability happens only if energy is available. In other words, instability is physically real only when the state from which the instability develops is thermodynamically out of equilibrium. In a system in thermodynamic equilibrium – which has no free energy – increasing solutions cannot exist.

9.2.2 The Fluid Approach

To illustrate our point let us take the simple example of the hydrodynamic **Rayleigh–Taylor** instability which will be studied in detail later. We have two fluids superimposed with the denser fluid above the less dense (i.e. with uniform densities such as $\rho_A > \rho_B$). Figure 9.2 shows this example with a non-rectilinear interface between the two fluids. The total (mechanical) energy of the system (which is assumed to be non-dissipative) is conserved. Any disturbance of the interface between the two fluids causes a decrease of the potential energy of the system and, therefore, an increase of the kinetic energy: the system is unstable.

In practice, one assumes the presence of a disturbance of the form $\exp(-i\omega t)$ in the fluid equations and one calculates the dispersion relation of the modes; one finds for this configuration

$$\omega_k^2 = -kg \left(\frac{\rho_A - \rho_B}{\rho_A + \rho_B} \right) , \tag{9.8}$$

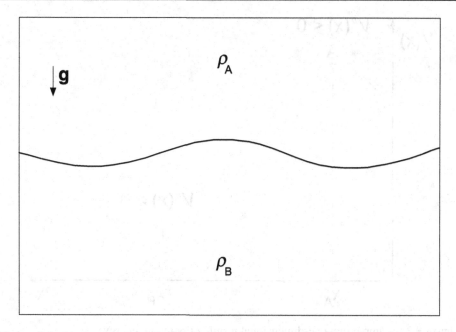

Figure 9.2 Configuration leading to the Rayleigh–Taylor instability if $\rho_A > \rho_B$.

where k is a wavenumber and g the gravitational acceleration. We see that if $\rho_A >$ ρ_B then $\omega_k^2 < 0$. In this case, the angular frequency is purely imaginary and the perturbation can grow exponentially.

9.3 The Energy Stability Criterion

9.3.1 A One-Dimensional Example

Consider the one-dimensional profile of Figure 9.3 for the potential energy. The force acting on the object studied is

$$F(x) = -\nabla V(x) = -\frac{dV}{dx}\mathbf{u_x}. \tag{9.9}$$

The positions x_A and x_B correspond to extrema A and B for which

$$F(x_i) = -\frac{dV}{dx} = 0. \tag{9.10}$$

These are equilibrium positions. The analysis in the vicinity of A and B gives us:

$$F(x) = F(x_i) + F'(x_i)(x - x_i) + \cdots$$
$$= 0 - V''(x_i)(x - x_i) + \cdots. \tag{9.11}$$

- At point A: $V''(x_i) < 0$, there is an unstable point.
- At point B: $V''(x_i) > 0$, there is a stable point.

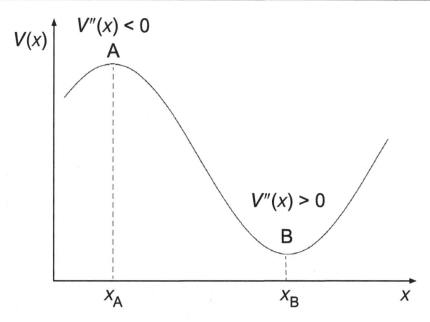

Figure 9.3 Example of a one-dimensional profile of potential energy.

We define $\xi = x - x_i$ and obtain approximately

$$m\frac{d^2\xi}{dt^2} = F'(x_i)\xi = -V''(x_i)\xi . \qquad (9.12)$$

This is a homogeneous second-order equation where the angular frequency ω is given by

$$\omega^2 = \frac{V''(x_i)}{m} = \frac{F'(x_i)}{m} . \qquad (9.13)$$

The solution of the differential equation is an oscillation at the frequency ω if $\omega^2 > 0$.

- At A, $\omega^2 < 0$ and thus $\omega = \pm i\sqrt{F'(x_i)/m}$. It is an **unstable** solution which increases exponentially in $\xi_i \exp\left(\sqrt{F'(x_i)/mt}\right)$.
- At B, $\omega^2 > 0$ and thus $\omega = \pm\sqrt{F'(x_i)/m}$. It is a **stable** solution oscillating around the equilibrium position.

Remark 1. We can already notice that ω^2 is still **real** if the system is globally **conservative**, like for example the inviscid and ideal MHD ($\nu = \eta = 0$).

Remark 2. If ω^2 is real, then the angular frequency ω is either purely real (pure oscillations) or purely imaginary (pure instability). Otherwise, we could have $\omega = \omega_r + i\omega_i$, i.e. damping or increasing oscillations. Consider the following example of Figure 9.4. We have $E = V + E_{cin}$ equal to a constant: the total energy of the system is conserved. In particular, we have $E_{cin}(x_{max}) = 0$ and $V(x_{max}) = E$. If the

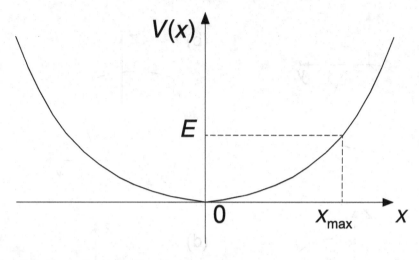

Figure 9.4 A one-dimensional well of potential energy.

motion is damped ($\omega_i < 0$) then x_{max} decreases over time and so does E, which is not possible according to the assumptions.

Remark 3. The result that has been discussed is related to the hermitian character of the small-displacement operator that will be introduced later.

9.3.2 Two-Dimensional Examples

Let us define $V(x, y)$, the potential energy of a two-dimensional system. The positions of equilibrium (stable or unstable) are determined by the conditions

$$\frac{\partial V}{\partial x} = \frac{\partial V}{\partial y} = 0. \tag{9.14}$$

Then, we have the four possible configurations shown in Figure 9.5.

- The first case (a) is an unstable equilibrium, for which

$$\frac{\partial^2 V}{\partial x^2} < 0, \quad \frac{\partial^2 V}{\partial y^2} < 0. \tag{9.15}$$

- The second case (b) is a stable equilibrium, for which

$$\frac{\partial^2 V}{\partial x^2} > 0, \quad \frac{\partial^2 V}{\partial y^2} > 0. \tag{9.16}$$

- The third case (c) is an unstable equilibrium, for which

$$\frac{\partial^2 V}{\partial x^2} > 0, \quad \frac{\partial^2 V}{\partial y^2} < 0. \tag{9.17}$$

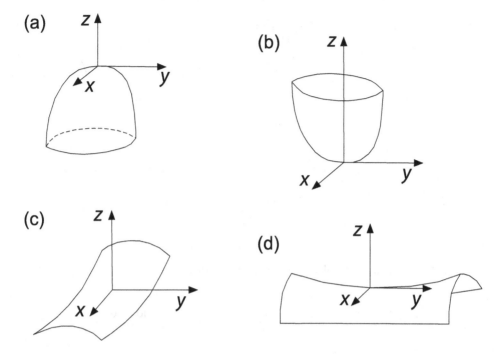

Figure 9.5 Examples of stable or unstable two-dimensional equilibrium.

- The last case (d) is also an unstable equilibrium, for which

$$\frac{\partial^2 V}{\partial x^2} < 0, \quad \frac{\partial^2 V}{\partial y^2} > 0. \tag{9.18}$$

It is thus seen that only 25% of two-dimensional equilibria are stable, whereas 50% of those in one dimension are stable. In general, it can be shown that in N dimensions, the proportion of stable equilibria is $1/2^N$. The number of stable equilibria therefore decreases with the number of space dimensions. In a plasma where the number of degrees of freedom $N \rightarrow +\infty$, if there is an equilibrium configuration this has plenty of chances to be unstable. It is therefore essential to find a general criterion of stability for a given physical situation. This is the subject of the present chapter.

9.3.3 The MHD Case

We recall that the standard inviscid and ideal MHD equations are

$$\frac{\partial \rho}{\partial t} + \nabla \cdot (\rho \mathbf{u}) = 0, \tag{9.19}$$

$$\rho \left(\frac{\partial \mathbf{u}}{\partial t} + \mathbf{u} \cdot \nabla \mathbf{u} \right) = -\nabla P + \mathbf{j} \times \mathbf{B} - \rho \nabla \Phi, \tag{9.20}$$

$$\frac{\partial \mathbf{B}}{\partial t} = \nabla \times (\mathbf{u} \times \mathbf{B}), \tag{9.21}$$

where Φ is the gravitational potential. Let us consider the closure equation,

$$\frac{d(P\rho^{-\gamma})}{dt} = 0, \tag{9.22}$$

with γ the polytropic index. In this case, the most general form of total energy conservation is

$$\frac{\partial}{\partial t} \left(\frac{\rho u^2}{2} + \frac{B^2}{2\mu_0} + \frac{P}{\gamma - 1} + \rho \Phi \right)$$

$$+ \nabla \cdot \left(\left(\frac{\rho u^2}{2} + \frac{\gamma P}{\gamma - 1} + \rho \Phi \right) \mathbf{u} + \mathbf{Q} + \frac{\mathbf{E} \times \mathbf{B}}{\mu_0} \right) = 0, \tag{9.23}$$

where we recognize in the first term the kinetic, magnetic, internal, and gravitational energies. The second term shows, in particular, the heat function $\mathbf{Q}$ and the Poynting vector. In all generality, we should add to the right a dissipative and a source term.

If one integrates the previous energy equation over all space (plasma plus vacuum), it simplifies because the term $\nabla \cdot (\,)$ gives a surface integral (Ostrogradsky's formula; see Appendix 2): the first three terms are zero outside the plasma and therefore disappear, and the last two terms also disappear if the system is isolated (volume $\mathcal{V}_\infty$). Hence

$$\iiint \left(\frac{\rho u^2}{2} + \frac{B^2}{2\mu_0} + \frac{P}{\gamma - 1} + \rho \Phi \right) d\mathcal{V} = E^{\mathrm{T}} = \text{constant}. \tag{9.24}$$

The first term on the left is the kinetic energy and the other three form the potential energy. Thus, we find

$$E_{\mathrm{cin}} + V = E^{\mathrm{T}}. \tag{9.25}$$

If for a perturbation of the system in equilibrium it is shown that $\delta V < 0$, then necessarily $\delta E_{\mathrm{cin}} > 0$ and the system is unstable against this perturbation (see the end of the next section).

9.4 Perturbation Theory

9.4.1 The Small-Displacement Operator

In this section, we present the formalism of the linear perturbation theory in the framework of the standard inviscid and ideal MHD which is an example of conservative system. The equation of motion is

$$\rho \left(\frac{\partial \mathbf{u}}{\partial t} + \mathbf{u} \cdot \nabla \mathbf{u} \right) = -\nabla P + \mathbf{j} \times \mathbf{B} - \rho \nabla \Phi, \tag{9.26}$$

where $\mathbf{g} = -\nabla\Phi_0$ is the gravitational acceleration. It is assumed that the system is initially in equilibrium (i.e. $\mathbf{u}(t = 0) = \mathbf{0}$), a situation for which we can write

$$\nabla P_0 = \frac{1}{\mu_0}(\nabla \times \mathbf{B}_0) \times \mathbf{B}_0 - \rho_0 \nabla\Phi_0 , \tag{9.27}$$

with $\mathbf{j}_0 = \nabla \times \mathbf{B}_0/\mu_0$. One adds a small perturbation to the system, which leads to the non-equilibrium equation of motion (for $t > 0$) at first order,

$$\rho_0 \dot{\mathbf{u}} = -\nabla P + \frac{1}{\mu_0}(\nabla \times \mathbf{B}) \times \mathbf{B} + \rho\mathbf{g} , \tag{9.28}$$

where one adopts the classical notation $\dot{\mathbf{u}} \equiv \partial\mathbf{u}/\partial t$. One may note that ρ_0 does not depend on time (according to the density equation). One gets, after differentiating with respect to time,

$$\rho_0 \ddot{\mathbf{u}} = -\nabla\dot{P} + \frac{1}{\mu_0}(\nabla \times \dot{\mathbf{B}}) \times \mathbf{B} + \frac{1}{\mu_0}(\nabla \times \mathbf{B}) \times \dot{\mathbf{B}} + \dot{\rho}\mathbf{g} . \tag{9.29}$$

The temporal derivatives of pressure, density, and magnetic field will be replaced with the terms which appear in the associated equations. The density equation is written at first order as

$$\dot{\rho} = -\nabla \cdot (\rho_0\mathbf{u}) . \tag{9.30}$$

The equation of state (with a polytropic closure) is

$$\frac{dP}{dt}\rho^{-\gamma} - \gamma\rho^{-(\gamma+1)}P\frac{d\rho}{dt} = 0 ,$$

$$\frac{dP}{dt} + \gamma P\nabla \cdot \mathbf{u} = 0 , \tag{9.31}$$

which gives at first order

$$\dot{P} + (\mathbf{u} \cdot \nabla)P_0 + \gamma P_0 \nabla \cdot \mathbf{u} = 0 . \tag{9.32}$$

Then Maxwell's equations and Ohm's law give us finally (at first order)

$$\dot{\mathbf{B}} = -\nabla \times \mathbf{E} = \nabla \times (\mathbf{u} \times \mathbf{B}_0) . \tag{9.33}$$

This allows us to rewrite the equation of motion **at first order** as

$$\rho_0\ddot{\mathbf{u}} = \nabla (\mathbf{u} \cdot \nabla P_0 + \gamma P_0 \nabla \cdot \mathbf{u}) + \frac{1}{\mu_0} (\nabla \times (\nabla \times (\mathbf{u} \times \mathbf{B}_0))) \times \mathbf{B}_0$$

$$+ \frac{1}{\mu_0}(\nabla \times \mathbf{B}_0) \times (\nabla \times (\mathbf{u} \times \mathbf{B}_0)) - \mathbf{g}(\nabla \cdot (\rho_0\mathbf{u})) . \tag{9.34}$$

We introduce now the **Lagrangian displacement** $\boldsymbol{\xi}(\mathbf{r}_0, t)$, where $\mathbf{r}_0$ is the initial position of equilibrium of the system. We have

$$\boxed{\mathbf{r} = \mathbf{r}_0 + \boldsymbol{\xi} ,} \tag{9.35}$$

with $\xi \ll r_0$. In this case

$$\mathbf{u}(\mathbf{r}, t) \simeq \mathbf{u}(\mathbf{r}_0, t) + (\xi \cdot \nabla)\mathbf{u}(\mathbf{r}_0, t) \simeq \mathbf{u}(\mathbf{r}_0, t) = \dot{\xi}(\mathbf{r}_0, t), \qquad (9.36)$$

$$\xi(\mathbf{r}_0, 0) = 0, \qquad (9.37)$$

$$\ddot{\xi}(\mathbf{r}_0, 0) = 0. \qquad (9.38)$$

The last two equations reflect the fact that the forces are zero in the initial position of equilibrium. Replacing this in the equation of motion and integrating it over t gives

$$\rho_0 \ddot{\xi} = \nabla(\xi \cdot \nabla P_0 + \gamma P_0 \nabla \cdot \xi) + \frac{1}{\mu_0}(\nabla \times (\nabla \times (\xi \times \mathbf{B}_0))) \times \mathbf{B}_0$$

$$+ \frac{1}{\mu_0}(\nabla \times \mathbf{B}_0) \times (\nabla \times (\xi \times \mathbf{B}_0)) - \mathbf{g}(\nabla \cdot (\rho_0 \xi)). \qquad (9.39)$$

We can write

$$\rho_0 \ddot{\xi}(\mathbf{r}_0, t) = \mathbf{F}(\xi(\mathbf{r}_0, t)), \qquad (9.40)$$

where $\mathbf{F}$ is the **small-displacement operator** (this is a linearized force density).

9.4.2 Solution to Initial Values

From the equilibrium configuration of the system $(\rho_0, P_0, \mathbf{B}_0)$ with the appropriate boundary conditions and the initial values of $\xi(\mathbf{r}_0, 0)$, $\dot{\xi}(\mathbf{r}_0, 0)$, and $\ddot{\xi}(\mathbf{r}_0, 0)$, Eq. (9.39) gives us $\xi(\mathbf{r}_0, t)$. Then, we can obtain the temporal evolution of the perturbed quantities:

$$\rho(\mathbf{r}_0, t) = \rho_0(\mathbf{r}_0) + \rho_1(\mathbf{r}_0, t), \qquad (9.41)$$

$$P(\mathbf{r}_0, t) = P_0(\mathbf{r}_0) + P_1(\mathbf{r}_0, t), \qquad (9.42)$$

$$\mathbf{B}(\mathbf{r}_0, t) = \mathbf{B}_0(\mathbf{r}_0) + \mathbf{B}_1(\mathbf{r}_0, t). \qquad (9.43)$$

To do this, we use

$$\dot{\mathbf{B}} = \nabla \times (\mathbf{u} \times \mathbf{B}) \simeq \nabla \times (\dot{\xi} \times \mathbf{B}_0) = \dot{\mathbf{B}}_1, \qquad (9.44)$$

hence

$$\mathbf{B}_1 = \nabla \times (\xi \times \mathbf{B}_0), \qquad (9.45)$$

which represents the local change of $\mathbf{B}$ (for δt). We have also

$$\mathbf{j} \times \mathbf{B} \simeq \frac{1}{\mu_0}(\nabla \times \mathbf{B}_1) \times \mathbf{B}_0 + \frac{1}{\mu_0}(\nabla \times \mathbf{B}_0) \times \mathbf{B}_1 + \frac{1}{\mu_0}(\nabla \times \mathbf{B}_0) \times \mathbf{B}_0 \quad (9.46)$$

and

$$\dot{P} = -\mathbf{u} \cdot \nabla P - \gamma P \nabla \cdot \mathbf{u} \simeq -\dot{\xi} \cdot \nabla P_0 - \gamma P_0 \nabla \cdot \dot{\xi} = \dot{P}_1, \qquad (9.47)$$

hence

$$P_1 = -\xi \cdot \nabla P_0 - \gamma P_0 \nabla \cdot \xi\,, \qquad (9.48)$$

which represents the local change of P (for δt).

9.4.3   The Equation of the Normal Modes

To solve the equation $\rho_0 \ddot{\xi} = \mathbf{F}(\xi)$, one assumes that one can separate the variables $\mathbf{r}_0$ and t in the following manner:

$$\xi(\mathbf{r}_0, t) = \xi(\mathbf{r}_0)\exp(-i\omega_k t)\,. \qquad (9.49)$$

The operator F being linear, we get

$$\mathbf{F}(\xi(\mathbf{r}_0, t)) = \mathbf{F}(\xi(\mathbf{r}_0))\exp(-i\omega_k t)\,. \qquad (9.50)$$

We can deduce the general equation for the eigenvalues,

$$\boxed{-\rho_0 \omega_k^2 \xi_k(\mathbf{r}_0) = \mathbf{F}(\xi_k(\mathbf{r}_0)),} \qquad (9.51)$$

with $(-\omega_k^2)$ the eigenvalues of the operator $\mathbf{F}/\rho_0$, $\xi_k(\mathbf{r}_0)$ the associated eigenvectors, and the general solution

$$\boxed{\xi(\mathbf{r}_0, t) = \sum_k a_k \exp(-i\omega_k t)\xi_k(\mathbf{r}_0).} \qquad (9.52)$$

We know that the values $\omega_k^2 < 0$ lead to unstable solutions which allow us to define the **stability criterion** of an MHD equilibrium as follows.

> *An MHD equilibrium is stable if and only if all the eigenvalues of the operator* $\mathbf{F}/\rho_0$ *are negative, i.e. if* $\omega_k^2 > 0$.

9.4.4   Properties of the Operator F

We have seen (admitted) that a physical system that is globally conservative has its eigenvalues, $-\omega_k^2$, real. This is related to the hermitian character of the small-displacement operator. We come back now to this property. Let us define

$$\mathbf{F}(\xi) = -\nabla V\,, \qquad (9.53)$$

hence

$$dV = \nabla V \cdot d\boldsymbol{\ell} = -\mathbf{F}(\boldsymbol{\ell}) \cdot d\boldsymbol{\ell}\,, \qquad (9.54)$$

and the variation of the potential energy is

$$\delta V = -\int_0^\xi \mathbf{F}(\boldsymbol{\ell}) \cdot d\boldsymbol{\ell} = -\frac{1}{2}\xi \cdot \mathbf{F}(\xi)\,. \qquad (9.55)$$

It should be noted that we have used the relations

$$\mathbf{F}(\boldsymbol{\xi}) = a_i \xi_i \mathbf{e_i} \,; \, \mathbf{F}(\boldsymbol{\ell}) \cdot d\boldsymbol{\ell} = a_i \ell_i \, d\ell_i \,; \, \int_0^{\boldsymbol{\xi}} a_i \ell_i \, d\ell_i = \frac{1}{2} a_i \xi_i^2 = \frac{1}{2} F_i \xi_i \,. \quad (9.56)$$

We obtain at first order

$$\frac{\partial}{\partial t} \iiint \left(\frac{1}{2} \rho_0 \dot{\xi}^2 - \frac{1}{2} \boldsymbol{\xi} \cdot \mathbf{F}(\boldsymbol{\xi}) \right) d\mathcal{V} = 0 \,,$$

$$\iiint \left(\rho_0 \dot{\boldsymbol{\xi}} \cdot \ddot{\boldsymbol{\xi}} - \frac{1}{2} \dot{\boldsymbol{\xi}} \cdot \mathbf{F}(\boldsymbol{\xi}) - \frac{1}{2} \boldsymbol{\xi} \cdot \mathbf{F}(\dot{\boldsymbol{\xi}}) \right) d\mathcal{V} = 0 \,, \quad (9.57)$$

$$\iiint \left(\frac{1}{2} \dot{\boldsymbol{\xi}} \cdot \mathbf{F}(\boldsymbol{\xi}) - \frac{1}{2} \boldsymbol{\xi} \cdot \mathbf{F}(\dot{\boldsymbol{\xi}}) \right) d\mathcal{V} = 0 \,,$$

and thus

$$\iiint \dot{\boldsymbol{\xi}} \cdot \mathbf{F}(\boldsymbol{\xi}) d\mathcal{V} = \iiint \boldsymbol{\xi} \cdot \mathbf{F}(\dot{\boldsymbol{\xi}}) d\mathcal{V} \,, \quad (9.58)$$

which demonstrates that the operator $\mathbf{F}$ is **hermitian**.

Let us go back to the relationship

$$- \rho_0 \omega_k^2 \boldsymbol{\xi}_k(\mathbf{r_0}) = \mathbf{F}(\boldsymbol{\xi}_k(\mathbf{r_0})) \,. \quad (9.59)$$

We multiply by $\boldsymbol{\xi}_k^*$ and integrate over the plasma volume $\mathcal{V}$; we get

$$- \omega_k^2 \iiint \rho_0 |\boldsymbol{\xi}_k|^2 \, d\mathcal{V} = \iiint \boldsymbol{\xi}_k^* \cdot \mathbf{F}(\boldsymbol{\xi}_k) d\mathcal{V}$$

$$(-\omega_k^2)^* \iiint \rho_0 |\boldsymbol{\xi}_k|^2 \, d\mathcal{V} = \iiint \boldsymbol{\xi}_k \cdot \mathbf{F}(\boldsymbol{\xi}_k^*) d\mathcal{V} \,. \quad (9.60)$$

By subtraction, we arrive at

$$(\omega_k^2 - (\omega_k^2)^*) \iiint \rho_0 |\boldsymbol{\xi}_k|^2 \, d\mathcal{V} = \iiint (\boldsymbol{\xi}_k \cdot \mathbf{F}(\boldsymbol{\xi}_k^*) - \boldsymbol{\xi}_k^* \cdot \mathbf{F}(\boldsymbol{\xi}_k)) d\mathcal{V} = 0 \,, \quad (9.61)$$

by hermiticity. Hence

$$\omega_k^2 = (\omega_k^2)^* \,. \quad (9.62)$$

Thus ω_k^2 is **purely real.**

Remark 4. We have simply checked a well-known theorem: "The eigenvalues of a hermitian operator are real".

Remark 5. There is another well-known theorem: "Eigenvectors associated with different eigenvalues of a hermitian operator are orthogonal".

Let us check this from the relation

$$- \rho_0 \omega_k^2 \boldsymbol{\xi}_k = \mathbf{F}(\boldsymbol{\xi}_k) \,, \quad (9.63)$$

hence

$$- \rho_0 \omega_k^2 \boldsymbol{\xi}_j \cdot \boldsymbol{\xi}_k = \boldsymbol{\xi}_j \cdot \mathbf{F}(\boldsymbol{\xi}_k) = \boldsymbol{\xi}_k \cdot \mathbf{F}(\boldsymbol{\xi}_j) = - \rho_0 \omega_j^2 \boldsymbol{\xi}_k \cdot \boldsymbol{\xi}_j,$$

$$(\omega_k^2 - \omega_j^2) \iiint \rho_0 \boldsymbol{\xi}_j \cdot \boldsymbol{\xi}_k \, d\mathcal{V} = \iiint \left(\boldsymbol{\xi}_k \cdot \mathbf{F}(\boldsymbol{\xi}_j) - \boldsymbol{\xi}_j \cdot \mathbf{F}(\boldsymbol{\xi}_k) \right) d\mathcal{V},$$

$$\iiint \rho_0 \boldsymbol{\xi}_j \cdot \boldsymbol{\xi}_k \, d\mathcal{V} = 0 \quad \text{if} \quad \omega_k^2 \neq \omega_j^2. \tag{9.64}$$

Therefore $\boldsymbol{\xi}_j$ and $\boldsymbol{\xi}_k$ are orthogonal. We will choose to write

$$\frac{1}{2} \iiint \rho_0 \boldsymbol{\xi}_j \cdot \boldsymbol{\xi}_k \, d\mathcal{V} = \delta_{kj}, \tag{9.65}$$

and the solution

$$\boldsymbol{\xi}(\mathbf{r}_0, t) = \sum_k a_k \exp(-i\omega_k t) \boldsymbol{\xi}_k(\mathbf{r}_0), \tag{9.66}$$

sets the **normal modes** $\boldsymbol{\xi}_k(\mathbf{r}_0)$ and normal coordinates a_k.

To conclude, we can notice that the problem with the eigenvalues we have just exposed is quite similar to that of quantum mechanics: the linearized MHD equations and the displacement vector are the analogues of the Schrödinger equation and the wave function.

9.4.5 The Return on the Energy Integral

In Section 9.3.3, we said that if $\delta V < 0$ then the system is unstable against the perturbation. Let us demonstrate this result. We have seen that

$$\delta V = -\frac{1}{2} \iiint \boldsymbol{\xi} \cdot \mathbf{F}(\boldsymbol{\xi}) d\mathcal{V} \tag{9.67}$$

is the variation of the total potential energy. From the above, we get

$$\delta V = -\frac{1}{2} \iiint \left(\sum_k a_k \exp(-i\omega_k t) \boldsymbol{\xi}_k(\mathbf{r}_0) \right) \cdot \mathbf{F} \left(\sum_\ell a_\ell \exp(-i\omega_\ell t) \boldsymbol{\xi}_\ell(\mathbf{r}_0) \right) d\mathcal{V}$$

$$= \frac{1}{2} \iiint \left(\sum_k a_k \exp(-i\omega_k t) \boldsymbol{\xi}_k(\mathbf{r}_0) \right) \cdot \left(\sum_\ell \rho_0 \omega_\ell^2 a_\ell \exp(-i\omega_\ell t) \boldsymbol{\xi}_\ell(\mathbf{r}_0) \right) d\mathcal{V}$$

$$= \frac{1}{2} \iiint \sum_{k,\ell} \omega_\ell^2 \rho_0 a_k a_\ell \exp(-i\omega_k t) \exp(-i\omega_\ell t) \, \boldsymbol{\xi}_k(\mathbf{r}_0) \cdot \boldsymbol{\xi}_\ell(\mathbf{r}_0) d\mathcal{V}$$

$$\sim \sum_k a_k^2 \omega_k^2 \cos^2(\omega_k t). \tag{9.68}$$

The last line is obtained by noting that $\boldsymbol{\xi}$ is real. As a result, δV is negative if at least an eigenvalue ω_k^2 is negative. Over time the instability grows and this term will dominate the others, then after a while we will have $\delta V < 0$.

10

Study of MHD Instabilities

The linear perturbation theory developed in Chapter 9 will be used in this chapter to solve several problems of instability in standard MHD. The first example that we will deal with is that of MHD waves for which we already know the solutions: the aim will be to see in practice how the formalism is applicable to simple problems. We will then study the Rayleigh–Taylor instability for neutral fluids and its MHD counterpart called the Schwarzschild–Kruskal instability where the magnetic field plays a stabilizing role. The z-pinch instability will then be discussed as an example of plasma confinement. Finally, we will conclude this chapter with the magneto-rotational instability, which is particularly relevant for accretion disks.

10.1 Stability of MHD Waves

We recall that the small-displacement equation in MHD (in the eigenvalue form) is

$$-\rho_0 \omega_k^2 \boldsymbol{\xi}_k(\mathbf{r}_0) = \mathbf{F}(\boldsymbol{\xi}_k(\mathbf{r}_0)),$$
(10.1)

with

$$\mathbf{F}(\boldsymbol{\xi}_k(\mathbf{r}_0)) = \nabla(\boldsymbol{\xi}_k \cdot \nabla P_0 + \gamma P_0 \nabla \cdot \boldsymbol{\xi}_k) + \frac{1}{\mu_0}\Big(\nabla \times (\nabla \times (\boldsymbol{\xi}_k \times \mathbf{B}_0))\Big) \times \mathbf{B}_0$$

$$+ \frac{1}{\mu_0}(\nabla \times \mathbf{B}_0) \times (\nabla \times (\boldsymbol{\xi}_k \times \mathbf{B}_0)) - \mathbf{g}(\nabla \cdot (\rho_0 \boldsymbol{\xi}_k)).$$
(10.2)

As a first application, we will consider an infinite homogeneous plasma imbedded in a uniform magnetic field (B_0 = constant) without gravity. Therefore P_0 = constant. We then obtain

$$\mathbf{F}(\boldsymbol{\xi}_k(\mathbf{r}_0)) = \gamma P_0 \nabla(\nabla \cdot \boldsymbol{\xi}_k) + \frac{1}{\mu_0}(\nabla \times (\nabla \times (\boldsymbol{\xi}_k \times \mathbf{B}_0))) \times \mathbf{B}_0$$

$$= \gamma P_0 \nabla(\nabla \cdot \boldsymbol{\xi}_k)$$

$$+ \frac{1}{\mu_0} \Big(\nabla \times \big(\xi_k (\nabla \cdot \mathbf{B}_0) + (\mathbf{B}_0 \cdot \nabla) \xi_k - \mathbf{B}_0 (\nabla \cdot \xi_k)$$

$$- (\xi_k \cdot \nabla) \mathbf{B}_0 \big) \Big) \times \mathbf{B}_0 , \tag{10.3}$$

and thus

$$\mathbf{F}(\xi_k(\mathbf{r}_0)) = \gamma P_0 \nabla (\nabla \cdot \xi_k) + \frac{1}{\mu_0} \Big(\nabla \times \big((\mathbf{B}_0 \cdot \nabla) \xi_k - \mathbf{B}_0 (\nabla \cdot \xi_k) \big) \Big) \times \mathbf{B}_0 .$$

10.1.1 Alfvén Waves

We will consider the incompressible case for which at first order

$$\nabla \cdot \dot{\xi}_k = 0 . \tag{10.4}$$

One assumes that $\nabla \cdot \xi_k = 0$ since initially the displacement is zero. Hence

$$-\rho_0 \omega_k^2 \xi_k = -\frac{\mathbf{B}_0}{\mu_0} \times (\nabla \times (\mathbf{B}_0 \cdot \nabla) \xi_k) . \tag{10.5}$$

One defines $\mathbf{B}_0 = B_0 \mathbf{e_z}$, with $\mathbf{e_z}$ a unit vector; one gets

$$\mu_0 \rho_0 \omega_k^2 \xi_k = B_0^2 \frac{\partial}{\partial z} (\mathbf{e_z} \times (\nabla \times \xi_k)) . \tag{10.6}$$

One has (by noticing that ξ_k is perpendicular to $\mathbf{e_z}$ after (10.6))

$$\mathbf{e_z} \times (\nabla \times \xi_k) = \mathbf{e_z} \times \left(\begin{pmatrix} \partial_x \\ \partial_y \\ \partial_z \end{pmatrix} \times \begin{pmatrix} \xi_x \\ \xi_y \\ 0 \end{pmatrix} \right)$$

$$= \mathbf{e_z} \times \begin{pmatrix} -\partial_z \xi_y \\ \partial_z \xi_x \\ \partial_x \xi_y - \partial_y \xi_x \end{pmatrix}$$

$$= -\frac{\partial \xi_x}{\partial z} \mathbf{e_x} - \frac{\partial \xi_y}{\partial z} \mathbf{e_y} = -\frac{\partial \xi_k}{\partial z} , \tag{10.7}$$

hence, finally,

$$\mu_0 \rho_0 \omega_k^2 \xi_k = -B_0^2 \frac{\partial^2 \xi_k}{\partial z^2} . \tag{10.8}$$

We seek a perturbative solution of the form $\xi_k = \xi_0 \exp(ikz)$. Hence the mode is given by

$$\boxed{\omega_k^2 = \frac{B_0^2 k^2}{\mu_0 \rho_0} > 0 .} \tag{10.9}$$

Thus, the system under consideration is **stable**. This solution is well known: it is an Alfvén wave that propagates along $\mathbf{B}_0$ with plasma displacements $\boldsymbol{\xi}_k$ perpendicular to $\mathbf{B}_0$ and a phase velocity

$$v_\phi = \frac{\omega_k}{k} = \frac{B_0}{\sqrt{\mu_0 \rho_0}} \equiv b_0 \tag{10.10}$$

equal to the Alfvén speed.

10.1.2   Magnetosonic Waves

We come back to the small-displacement equation

$$-\rho_0 \omega_k^2 \boldsymbol{\xi}_k(\mathbf{r}_0) = \mathbf{F}(\boldsymbol{\xi}_k(\mathbf{r}_0)), \tag{10.11}$$

with

$$\mathbf{F}(\boldsymbol{\xi}_k) = \gamma P_0 \nabla(\nabla \cdot \boldsymbol{\xi}_k) + \frac{1}{\mu_0}(\nabla \times (\nabla \times (\boldsymbol{\xi}_k \times \mathbf{B}_0))) \times \mathbf{B}_0. \tag{10.12}$$

One seeks perturbative solutions of the form $\boldsymbol{\xi}_k = \boldsymbol{\xi}_0 \exp(ikz)$. Hence

$$-\rho_0 \omega_k^2 \boldsymbol{\xi}_0 = -\gamma P_0 \mathbf{k}(\mathbf{k} \cdot \boldsymbol{\xi}_0) - \frac{1}{\mu_0}(\mathbf{k} \times (\mathbf{k} \times (\boldsymbol{\xi}_0 \times \mathbf{B}_0))) \times \mathbf{B}_0, \tag{10.13}$$

or

$$\rho_0 \omega_k^2 \boldsymbol{\xi}_0 = \gamma P_0 \mathbf{k}(\mathbf{k} \cdot \boldsymbol{\xi}_0) + \frac{1}{\mu_0}(\mathbf{k} \times ((\mathbf{k} \cdot \mathbf{B}_0)\boldsymbol{\xi}_0 - (\mathbf{k} \cdot \boldsymbol{\xi}_0)\mathbf{B}_0)) \times \mathbf{B}_0. \tag{10.14}$$

- If $\mathbf{k} \parallel \mathbf{B}_0$ (parallel propagation) one has

$$\mathbf{k} \times \mathbf{B}_0 = \mathbf{0}, \tag{10.15}$$

and

$$(\mathbf{k} \times (\mathbf{k} \cdot \mathbf{B}_0)\boldsymbol{\xi}_0) \times \mathbf{B}_0 = (\mathbf{k} \cdot \mathbf{B}_0)^2 \boldsymbol{\xi}_0 - (\mathbf{k} \cdot \mathbf{B}_0)(\boldsymbol{\xi}_0 \cdot \mathbf{B}_0)\mathbf{k}, \tag{10.16}$$

hence the relation

$$\left(\rho_0 \omega_k^2 - \frac{(\mathbf{k} \cdot \mathbf{B}_0)^2}{\mu_0}\right) \boldsymbol{\xi}_0 = \left(\gamma P_0(\mathbf{k} \cdot \boldsymbol{\xi}_0) - \frac{(\mathbf{k} \cdot \mathbf{B}_0)(\boldsymbol{\xi}_0 \cdot \mathbf{B}_0)}{\mu_0}\right) \mathbf{k}. \tag{10.17}$$

In the case $\boldsymbol{\xi}_0 \parallel \mathbf{k}$

$$\rho_0 \omega_k^2 \boldsymbol{\xi}_0 = \gamma P_0(\mathbf{k} \cdot \boldsymbol{\xi}_0)\mathbf{k} = \gamma P_0 k^2 \boldsymbol{\xi}_0, \tag{10.18}$$

and finally

$$\boxed{\omega_k^2 = \frac{\gamma P_0 k^2}{\rho_0} > 0.} \tag{10.19}$$

This **stable** mode is the sonic wave whose phase velocity is

$$v_\phi = \frac{\omega_k}{k} = \sqrt{\frac{\gamma P_0}{\rho_0}} \equiv c_S . \tag{10.20}$$

If $\boldsymbol{\xi}_0$ and $\mathbf{k}$ **are not** parallel, the coefficients of $\boldsymbol{\xi}_0$ and $\mathbf{k}$ in expression (10.17) must cancel out. This yields on the one hand

$$\rho_0 \omega_k^2 = \frac{k^2 B_0^2}{\mu_0} , \tag{10.21}$$

which is the Alfvén wave for which $\omega_k^2 > 0$, and on the other hand

$$\gamma P_0 (\mathbf{k} \cdot \boldsymbol{\xi}_0) = \frac{(\mathbf{k} \cdot \mathbf{B}_0)(\boldsymbol{\xi}_0 \cdot \mathbf{B}_0)}{\mu_0} = \frac{k B_0 (\boldsymbol{\xi}_0 \cdot \mathbf{B}_0)}{\mu_0} = \frac{B_0^2 (\boldsymbol{\xi}_0 \cdot \mathbf{k})}{\mu_0} . \tag{10.22}$$

This implies that $\mathbf{k} \perp \boldsymbol{\xi}_0$.
- If $\mathbf{k} \perp \mathbf{B}_0$ (perpendicular propagation) one has

$$\mathbf{k} \cdot \mathbf{B}_0 = 0 , \tag{10.23}$$

and

$$\begin{aligned}
\rho_0 \omega_k^2 \boldsymbol{\xi}_0 &= \gamma P_0 \mathbf{k}(\mathbf{k} \cdot \boldsymbol{\xi}_0) - \frac{1}{\mu_0} (\mathbf{k} \times ((\mathbf{k} \cdot \boldsymbol{\xi}_0)\mathbf{B}_0)) \times \mathbf{B}_0 \\
&= \gamma P_0 \mathbf{k}(\mathbf{k} \cdot \boldsymbol{\xi}_0) - \frac{1}{\mu_0}(\mathbf{k} \cdot \mathbf{B}_0)(\mathbf{k} \cdot \boldsymbol{\xi}_0)\mathbf{B}_0 + \frac{1}{\mu_0}(\mathbf{k} \cdot \boldsymbol{\xi}_0)B_0^2 \mathbf{k} \\
&= \gamma P_0 \mathbf{k}(\mathbf{k} \cdot \boldsymbol{\xi}_0) + \frac{1}{\mu_0}(\mathbf{k} \cdot \boldsymbol{\xi}_0)B_0^2 \mathbf{k} ,
\end{aligned} \tag{10.24}$$

for which the only solution is $\boldsymbol{\xi}_0 \parallel \mathbf{k}$. Hence the relation

$$\rho_0 \omega_k^2 = \gamma P_0 k^2 + \frac{B_0^2}{\mu_0} k^2 , \tag{10.25}$$

and finally

$$\boxed{\omega_k^2 = \left(\frac{\gamma P_0}{\rho_0} + \frac{B_0^2}{\rho_0 \mu_0}\right) k^2 = (c_S^2 + b_0^2)k^2 > 0 .} \tag{10.26}$$

We still find a **stable** solution: this is the magnetosonic wave whose phase velocity is

$$v_\phi = \frac{\omega_k}{k} = \sqrt{c_S^2 + b_0^2} . \tag{10.27}$$

One can also solve the small-displacement equation in the general case (see also Chapter 4) for an arbitrary direction $\mathbf{k}$ compared with $\mathbf{B}_0$.

10.2   Rayleigh–Taylor Instability

We will study the Rayleigh–Taylor instability by both methods introduced in the previous chapter (energy integrals and normal modes). We consider two fluids of uniform density ρ_A and ρ_B, with the denser fluid above. Figure 10.1 shows this example with a non-rectilinear interface between the two fluids. We consider the hydrodynamic incompressible case ($B = 0$).

10.2.1   The First Method: Energy Integrals

We start from the small-displacement equation

$$\rho_0 \ddot{\xi} = \mathbf{F}(\xi) , \tag{10.28}$$

with

$$\mathbf{F}(\xi) = -\nabla P_1 - \mathbf{g} \nabla \cdot (\rho_0 \xi) , \tag{10.29}$$

and, according to relation (9.48),

$$P_1 = -\xi \cdot \nabla P_0 - \gamma P_0 \nabla \cdot \xi = -\xi \cdot \nabla P_0 . \tag{10.30}$$

One notices that

$$\nabla \cdot (\rho_0 \xi) = \xi \cdot \nabla \rho_0 + \rho_0 \nabla \cdot \xi = \xi \cdot \nabla \rho_0 . \tag{10.31}$$

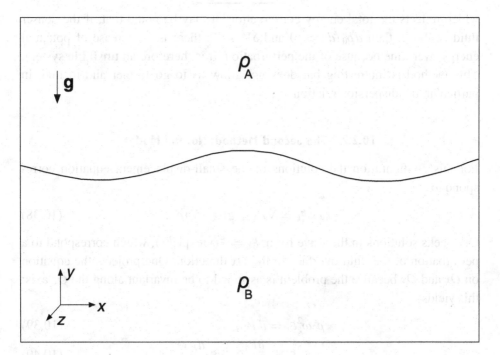

Figure 10.1 A configuration leading to Rayleigh–Taylor instability if $\rho_A > \rho_B$.

After the multiplication by $\dot{\xi}$ and integration over the volume containing the fluids, one gets

$$\frac{1}{2}\iiint \rho_0 \frac{\partial (\dot{\xi})^2}{\partial t}\, dV = -\iiint \dot{\xi} \cdot [\nabla P_1 + \mathbf{g}(\xi \cdot \nabla \rho_0)]dV. \qquad (10.32)$$

Since

$$\nabla \cdot (P_1\dot{\xi}) = P_1\nabla \cdot \dot{\xi} + \dot{\xi} \cdot \nabla P_1 = \dot{\xi} \cdot \nabla P_1, \qquad (10.33)$$

$$\dot{\xi} \cdot \mathbf{g}(\xi \cdot \nabla \rho_0) = -\dot{\xi}_y g \xi_y \frac{d\rho_0}{dy} = -\frac{1}{2}\frac{\partial}{\partial t}\left(\xi_y^2 g \frac{d\rho_0}{dy}\right), \qquad (10.34)$$

we have

$$\frac{1}{2}\iiint \rho_0 \frac{\partial (\dot{\xi})^2}{\partial t}\, dV = \iiint [-\nabla \cdot (P_1\dot{\xi})]dV + \frac{1}{2}\iiint \frac{\partial}{\partial t}\left[\xi_y^2 g \frac{d\rho_0}{dy}\right]dV. \qquad (10.35)$$

The first term on the right-hand side is zero because the normal component of the vector $\dot{\xi}$ to the rigid surface surrounding the fluids is by definition zero. By integrating over time we obtain

$$\frac{1}{2}\iiint \rho_0 (\dot{\xi})^2\, dV - \frac{1}{2}\iiint \xi_y^2 g \frac{d\rho_0}{dy}\, dV = 0; \qquad (10.36)$$

in other words

$$\boxed{\delta E_{\text{cin}} + \delta V = 0,} \qquad (10.37)$$

which reflects the total energy conservation. It may be noted that, if the denser fluid is above, then $d\rho_0/dy > 0$ and $\delta V < 0$: there is a decrease of potential energy over time because of the perturbation. It is therefore an **unstable** system. This method is interesting but does not allow us to go further and to find, in particular, the dispersion relation.

10.2.2 The Second Method: Normal Modes

For this configuration the solutions of the small-displacement equation correspond to

$$\rho_0\omega_k^2\xi_k = \nabla P_1 + \mathbf{g}(\xi_k \cdot \nabla)\rho_0. \qquad (10.38)$$

One seeks solutions in the wave form, $\xi_k = \xi(y)\exp(ikx)$, which correspond to a perturbation of the fluid interface in the Ox direction. One projects the equation on Ox and Oy because the problem is assumed to be invariant along the Oz axis; this yields

$$\rho_0\omega_k^2\xi_x = ikP_1, \qquad (10.39)$$

$$\rho_0\omega_k^2\xi_y = \frac{\partial P_1}{\partial y} - g\xi_y\frac{d\rho_0}{dy}. \qquad (10.40)$$

This gives

$$\rho_0 \omega_k^2 \xi_y = \frac{\omega_k^2}{ik} \frac{\partial}{\partial y}(\rho_0 \xi_x) - g\xi_y \frac{d\rho_0}{dy}. \tag{10.41}$$

After differentiating in the x direction, one has

$$ik\rho_0 \omega_k^2 \xi_y = \frac{\omega_k^2}{ik} \frac{\partial}{\partial y} \frac{\partial}{\partial x}(\rho_0 \xi_x) - gik\xi_y \frac{d\rho_0}{dy}. \tag{10.42}$$

Relation $\nabla \cdot \boldsymbol{\xi} = 0$ can be written

$$\frac{\partial \xi_x}{\partial x} + \frac{\partial \xi_y}{\partial y} = 0, \tag{10.43}$$

hence

$$ik\rho_0 \omega_k^2 \xi_y = -\frac{\omega_k^2}{ik} \frac{\partial}{\partial y} \left(\rho_0 \frac{\partial \xi_y}{\partial y} \right) - gik\xi_y \frac{d\rho_0}{dy}, \tag{10.44}$$

and finally

$$\boxed{k^2 \left(\rho_0 \omega_k^2 + g\frac{d\rho_0}{dy} \right) \xi_y = \omega_k^2 \frac{\partial}{\partial y} \left(\rho_0 \frac{\partial \xi_y}{\partial y} \right).} \tag{10.45}$$

In the ideal situation that we consider there is a jump of density at the interface between the two fluids and a uniform density within each of the fluids. We can deduce the following two cases.

- For $y > 0$ or $y < 0$ (far from the interface),

$$\frac{d\rho_0}{dy} = 0, \tag{10.46}$$

and thus relation (10.45) simplifies to

$$k^2 \xi_y = \frac{\partial^2 \xi_y}{\partial y^2}. \tag{10.47}$$

Hence, the solution is

$$\xi_y = \xi_{y0} \exp(-k|y|)\exp(ikx). \tag{10.48}$$

This solution is chosen so that $\xi_y \to 0$ when $|y| \to \infty$.

- In the vicinity of the interface,

$$\rho_0 = \rho_A \theta(y) + \rho_B \theta(-y), \tag{10.49}$$

where θ is the Heaviside function, namely

$$\theta = \begin{cases} 1, & \text{if } y > 0, \\ 0, & \text{otherwise.} \end{cases} \tag{10.50}$$

Hence

$$\frac{d\rho_0}{dy}\bigg|_0 = \rho_A \delta(y) - \rho_B \delta(y) \tag{10.51}$$

After integration of expression (10.45) over y between $-\varepsilon$ and $+\varepsilon$, one obtains

$$k^2 \omega_k^2 \int_{-\varepsilon}^{+\varepsilon} [\rho_A \theta(y) + \rho_B \theta(-y)] \xi_y \, dy + k^2 g \int_{-\varepsilon}^{+\varepsilon} [\rho_A \delta(y) - \rho_B \delta(y)] \xi_y \, dy$$

$$= \omega_k^2 \int_{-\varepsilon}^{+\varepsilon} \frac{\partial}{\partial y} \left(\rho_0 \frac{\partial \xi_y}{\partial y} \right) dy, \tag{10.52}$$

then, noticing that ξ_y is continuous at the interface, one finds

$$k^2 \omega_k^2 \left(\rho_A \int_0^{+\varepsilon} \xi_y \, dy + \rho_B \int_{-\varepsilon}^0 \xi_y \, dy \right) + k^2 g (\rho_A - \rho_B) \xi_y(0)$$

$$= \omega_k^2 \left(\rho_A \frac{\partial \xi_y}{\partial y} \bigg|_\varepsilon - \rho_B \frac{\partial \xi_y}{\partial y} \bigg|_{-\varepsilon} \right) = \omega_k^2 \left(-\rho_A k \xi_y - \rho_B k \xi_y \right). \tag{10.53}$$

Hence

$$k^2 g (\rho_A - \rho_B) \xi_y(0) = -\omega_k^2 k \xi_y(0)(\rho_A + \rho_B), \tag{10.54}$$

and finally the dispersion relation of the perturbed system is given by

$$\boxed{\omega_k^2 = -kg \left(\frac{\rho_A - \rho_B}{\rho_A + \rho_B} \right).} \tag{10.55}$$

The system is therefore **unstable** when $\rho_A > \rho_B$ since then $\omega_k^2 < 0$. We see that the instability is growing even faster than with k: in other words, the instability preferentially grows at small scale. However, if $\rho_A < \rho_B$, the system is unconditionally stable: one has gravity waves at the interface between the two fluids. Figure 10.2 shows the Rayleigh–Taylor instability: this is a numerical simulation in which the heavier fluid is initially at the top and the lightest at the bottom.

10.3 Kruskal–Schwarzschild Instability

Let us consider the previous configuration (incompressible case) and add a magnetic field in the Oz direction as shown in Figure 10.3. This is a configuration which can lead to the **Kruskal–Schwarzschild** instability (Kruskal and Schwarzschild, 1954). We will demonstrate that the magnetic field has a stabilizing role regarding the perturbation. To do this, we directly apply the

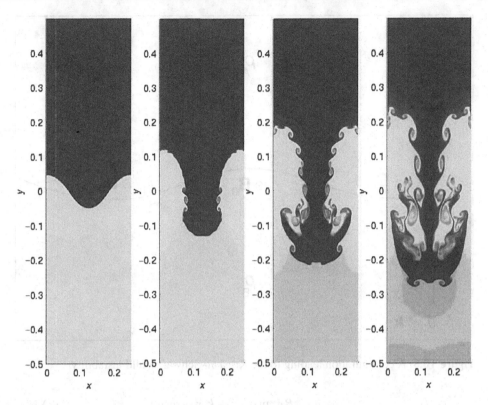

Figure 10.2 Hydrodynamic simulation of the Rayleigh–Taylor instability: the heavier fluid (dark part) dives into the lighter (clear part) and generates eddies. Image in the public domain.

method of normal modes. One starts from the incompressible small-displacement equation:

$$\rho_0 \omega_k^2 \boldsymbol{\xi}_k = \nabla P_1 + \mathbf{g}(\boldsymbol{\xi}_k \cdot \nabla)\rho_0 - \frac{1}{\mu_0}(\nabla \times (\nabla \times (\boldsymbol{\xi}_k \times \mathbf{B}_0))) \times \mathbf{B}_0$$

$$- \frac{1}{\mu_0}(\nabla \times \mathbf{B}_0) \times (\nabla \times (\boldsymbol{\xi}_k \times \mathbf{B}_0)), \tag{10.56}$$

with $\mathbf{B}_0 = B_0(y)\mathbf{e}_z$ a magnetic field localized around the interface. One considers a perturbation of the form

$$\boldsymbol{\xi}_k = \boldsymbol{\xi}(y)\exp(i(zk\cos\theta + xk\sin\theta - \omega t))$$

$$= \boldsymbol{\xi}(y)\exp(i(zk_z + xk_x - \omega t)). \tag{10.57}$$

One obtains

$$\nabla \times (\boldsymbol{\xi}_k \times \mathbf{B}) = \nabla \times \begin{pmatrix} \xi_y B \exp(i(k_z z + k_x x - \omega t)) \\ -\xi_x B \exp(i(k_z z + k_x x - \omega t)) \\ 0 \end{pmatrix}$$

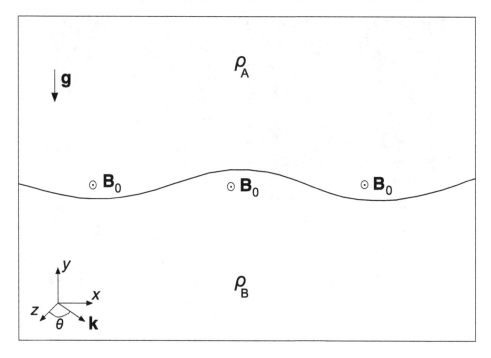

Figure 10.3 A configuration leading to the Kruskal–Schwarzschild instability.

$$= \begin{pmatrix} \partial_z(\xi_x B \exp(i(k_z z + k_x x - \omega t))) \\ \partial_z(\xi_y B \exp(i(k_z z + k_x x - \omega t))) \\ -\partial_x(\xi_x B \exp(i(k_z z + k_x x - \omega t))) - \partial_y(\xi_y B \exp(i(k_z z + k_x x - \omega t))) \end{pmatrix}$$

$$= \begin{pmatrix} \xi_x B i k_z \exp(i(k_z z + k_x x - \omega t)) \\ \xi_y B i k_z \exp(i(k_z z + k_x x - \omega t)) \\ -\xi_x B i k_x \exp(i(k_z z + k_x x - \omega t)) - \partial_y(\xi_y B)\exp(i(k_z z + k_x x - \omega t)) \end{pmatrix}.$$

$$\tag{10.58}$$

The incompressibility condition corresponds to

$$ik_x \xi_x + \partial_y \xi_y + ik_z \xi_z = 0. \tag{10.59}$$

This gives

$$\mathbf{B}_1 = \nabla \times (\boldsymbol{\xi}_k \times \mathbf{B})$$
$$= \begin{pmatrix} \xi_x B i k_z \exp(i(k_z z + k_x x - \omega t)) \\ \xi_y B i k_z \exp(i(k_z z + k_x x - \omega t)) \\ (-\xi_y \, \partial_y B + \xi_z B i k_z)\exp(i(k_z z + k_x x - \omega t)) \end{pmatrix}$$
$$= B i k_z \boldsymbol{\xi}_k - \xi_y \, \partial_y B \mathbf{e_z}. \tag{10.60}$$

On the other hand, we have the following vectorial equation:

$$(\nabla \times \mathbf{B}_1) \times \mathbf{B}_0 + (\nabla \times \mathbf{B}_0) \times \mathbf{B}_1$$
$$= -\nabla(\mathbf{B}_1 \cdot \mathbf{B}_0) + (\mathbf{B}_0 \cdot \nabla)\mathbf{B}_1 + (\mathbf{B}_1 \cdot \nabla)\mathbf{B}_0. \tag{10.61}$$

Hence

$$\rho_0\omega_k^2\boldsymbol{\xi}_k = \nabla\left(P_1 + \frac{\mathbf{B}_1 \cdot \mathbf{B}_0}{\mu_0}\right) + \mathbf{g}(\boldsymbol{\xi}_k \cdot \nabla)\rho_0 - \frac{1}{\mu_0}((\mathbf{B}_0 \cdot \nabla)\mathbf{B}_1 + (\mathbf{B}_1 \cdot \nabla)\mathbf{B}_0)$$

$$= \nabla\left(P_1 + \frac{\mathbf{B}_1 \cdot \mathbf{B}_0}{\mu_0}\right) + \mathbf{g}(\boldsymbol{\xi}_k \cdot \nabla)\rho_0 - \frac{1}{\mu_0}(B_0 \partial_z \mathbf{B}_1 + B_{1y} \partial_y \mathbf{B}_0)$$

$$= \nabla\left(P_1 + \frac{\mathbf{B}_1 \cdot \mathbf{B}_0}{\mu_0}\right) + \mathbf{g}(\boldsymbol{\xi}_k \cdot \nabla)\rho_0 + \frac{1}{\mu_0}B_0^2 k_z^2 \boldsymbol{\xi}_k. \tag{10.62}$$

For each component, one gets

$$\rho_0\omega_k^2\xi_x = ik_x\left(P_1 + \frac{\mathbf{B}_1 \cdot \mathbf{B}_0}{\mu_0}\right) + \frac{1}{\mu_0}B_0^2 k_z^2 \xi_x, \tag{10.63}$$

$$\rho_0\omega_k^2\xi_y = \partial_y\left(P_1 + \frac{\mathbf{B}_1 \cdot \mathbf{B}_0}{\mu_0}\right) - g\xi_y\frac{d\rho_0}{dy} + \frac{1}{\mu_0}B_0^2 k_z^2 \xi_y, \tag{10.64}$$

$$\rho_0\omega_k^2\xi_z = ik_z\left(P_1 + \frac{\mathbf{B}_1 \cdot \mathbf{B}_0}{\mu_0}\right) + \frac{1}{\mu_0}B_0^2 k_z^2 \xi_z, \tag{10.65}$$

or

$$\left(\rho_0\omega_k^2 - \frac{(\mathbf{B}_0 \cdot \mathbf{k})^2}{\mu_0}\right)\xi_x = ik_x\left(P_1 + \frac{\mathbf{B}_1 \cdot \mathbf{B}_0}{\mu_0}\right), \tag{10.66}$$

$$\left(\rho_0\omega_k^2 - \frac{(\mathbf{B}_0 \cdot \mathbf{k})^2}{\mu_0} + g\frac{d\rho_0}{dy}\right)\xi_y = \partial_y\left(P_1 + \frac{\mathbf{B}_1 \cdot \mathbf{B}_0}{\mu_0}\right), \tag{10.67}$$

$$\left(\rho_0\omega_k^2 - \frac{(\mathbf{B}_0 \cdot \mathbf{k})^2}{\mu_0}\right)\xi_z = ik_z\left(P_1 + \frac{\mathbf{B}_1 \cdot \mathbf{B}_0}{\mu_0}\right). \tag{10.68}$$

One inserts the first expression into the second. This yields

$$\left(\rho_0\omega_k^2 - \frac{(\mathbf{B}_0 \cdot \mathbf{k})^2}{\mu_0} + g\frac{d\rho_0}{dy}\right)\xi_y = \partial_y\left(\frac{\rho_0\omega_k^2 - (\mathbf{B}_0 \cdot \mathbf{k})^2/\mu_0}{ik_x}\xi_x\right), \tag{10.69}$$

and one derives in x

$$ik_x\left(\rho_0\omega_k^2 - \frac{(\mathbf{B}_0 \cdot \mathbf{k})^2}{\mu_0} + g\frac{d\rho_0}{dy}\right)\xi_y = \partial_y\left(\frac{\rho_0\omega_k^2 - (\mathbf{B}_0 \cdot \mathbf{k})^2/\mu_0}{ik_x}\partial_x\xi_x\right). \tag{10.70}$$

The same manipulation with the z component gives

$$\left(\rho_0\omega_k^2 - \frac{(\mathbf{B}_0 \cdot \mathbf{k})^2}{\mu_0} + g\frac{d\rho_0}{dy}\right)\xi_y = \partial_y\left(\frac{\rho_0\omega_k^2 - (\mathbf{B}_0 \cdot \mathbf{k})^2/\mu_0}{ik_z}\xi_z\right), \tag{10.71}$$

and, after a differentiation with respect to z,

$$ik_z\left(\rho_0\omega_k^2 - \frac{(\mathbf{B}_0 \cdot \mathbf{k})^2}{\mu_0} + g\frac{d\rho_0}{dy}\right)\xi_y = \partial_y\left(\frac{\rho_0\omega_k^2 - (\mathbf{B}_0 \cdot \mathbf{k})^2/\mu_0}{ik_z}\partial_z\xi_z\right). \tag{10.72}$$

We add the two expressions found after multiplication by ik_x and ik_z, respectively, which gives finally

$$k^2 \left(\rho_0 \omega_k^2 - \frac{(\mathbf{B}_0 \cdot \mathbf{k})^2}{\mu_0} + g \frac{d\rho_0}{dy} \right) \xi_y = \partial_y \left(\left(\rho_0 \omega_k^2 - \frac{(\mathbf{B}_0 \cdot \mathbf{k})^2}{\mu_0} \right) \partial_y \xi_y \right).$$

(10.73)

The final solution is obtained by considering the following two cases.

- For $y > 0$ or $y < 0$ (far from the interface),

$$\frac{d\rho_0}{dy} = 0,$$

(10.74)

$$\frac{dB_0}{dy} = 0,$$

(10.75)

hence

$$\partial_y^2 \xi_y - k^2 \xi_y = 0,$$

(10.76)

whose solution can trivially be written as

$$\xi_y \sim \exp(-k|y|).$$

(10.77)

- In the **vicinity** of the interface,

$$\rho_0 = \rho_A \theta(y) + \rho_B \theta(-y),$$

(10.78)

and thus

$$\left. \frac{d\rho_0}{dy} \right|_0 = \rho_A \delta(y) - \rho_B \delta(y).$$

(10.79)

After integration of expression (10.73) around the interface, one obtains

$$k^2 \omega_k^2 \int_{-\varepsilon}^{+\varepsilon} (\rho_A \theta(y) + \rho_B \theta(-y)) \xi_y \, dy - \frac{k^2}{\mu_0} \int_{-\varepsilon}^{+\varepsilon} (\mathbf{B}_0 \cdot \mathbf{k})^2 \xi_y \, dy$$

$$+ k^2 g \int_{-\varepsilon}^{+\varepsilon} (\rho_A \delta(y) - \rho_B \delta(y)) \xi_y \, dy = \left[\left(\rho_0 \omega_k^2 - \frac{(\mathbf{B}_0 \cdot \mathbf{k})^2}{\mu_0} \right) \partial_y \xi_y \right]_{-\varepsilon}^{+\varepsilon},$$

(10.80)

or, with the solution (10.77),

$$k^2 \omega_k^2 \left(\rho_A \int_0^{+\varepsilon} \xi_y \, dy + \rho_B \int_{-\varepsilon}^0 \xi_y \, dy \right) - \frac{k^2}{\mu_0} \int_{-\varepsilon}^{+\varepsilon} (\mathbf{B}_0 \cdot \mathbf{k})^2 \xi_y \, dy$$

$$+ k^2 g (\rho_A - \rho_B) \xi_y(0) =$$

$$- \left[\rho_A \omega_k^2 - \frac{(\mathbf{B}_0 \cdot \mathbf{k})^2}{\mu_0} \right] k \xi_y(0) - \left[\rho_B \omega_k^2 - \frac{(\mathbf{B}_0 \cdot \mathbf{k})^2}{\mu_0} \right] k \xi_y(0).$$

(10.81)

The three integrals tend to zero when $\varepsilon \to 0$, hence finally

$$\omega_k^2 = -kg\left(\frac{\rho_A - \rho_B}{\rho_A + \rho_B}\right) + \frac{2}{\mu_0}\frac{(\mathbf{B}_0 \cdot \mathbf{k})^2}{\rho_A + \rho_B}. \tag{10.82}$$

We see that in all cases where $\rho_A < \rho_B$ the system is stable. As in the purely hydrodynamic case (Rayleigh–Taylor instability), the first term can be destabilized when $\rho_A > \rho_B$ and its action is even stronger the larger k is. On the contrary, the second term – magnetic by nature – is **stabilizing**: the magnetic tension which appears at the interface helps to maintain the fluid above. However, this tension becomes zero for $\mathbf{k} \perp \mathbf{B}_0$.

Figure 10.4 shows the example of solar filaments (or protuberances), which can be interpreted as a plasma in suspension above the solar surface: the magnetic field plays the role of a magnetic mattress. These filaments appear dark on these images obtained in Hα at 656 nm with the spectroheliograph of the Paris–Meudon Observatory.

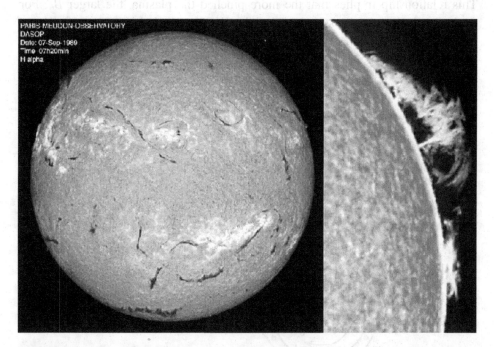

Figure 10.4 Left: the dark filaments are magnetized structures in the solar corona above the solar surface (image in Hα at 656 nm obtained by the spectroheliograph of the Paris–Meudon Observatory). Their height can reach 50 000 km and their length several hundred thousand kilometers. Their temperature is 100 times lower than that of the solar corona and their density at least a hundred times greater. Right: the same structures seen at the edge of the Sun. Credits: LESIA/Observatoire de Paris.

10.4 z-Pinch Instability

We shall study the cylindrical configuration of a z-pinch with an electric current in the z direction. The configuration differs slightly from that presented in Chapter 8 in the sense that a strong ring current is produced directly in the plasma (see Figure 10.5). This configuration was studied in the 1950s mainly in the framework of thermonuclear plasmas: the objective was to find a way to confine a compressible plasma by application of a magnetic field. We will demonstrate that unfortunately this simple configuration cannot be stable when the magnetic field is only azimuthal.

10.4.1   Static Equilibrium

A first general result can be shown simply from Ampère's theorem (see Figure 10.7 later):

$$\oint_\Gamma \mathbf{B} \cdot d\boldsymbol{\ell} = \oint_\Gamma B_\theta \, d\ell = \iint_S (\nabla \times \mathbf{B}) \cdot d\mathbf{S} = \mu_0 \iint_S \mathbf{j} \cdot d\mathbf{S} = \mu_0 I . \qquad (10.83)$$

This relationship implies that the more pinched the plasma, the larger B_θ. For further analysis, it is necessary to determine the pressure balance from the plasma

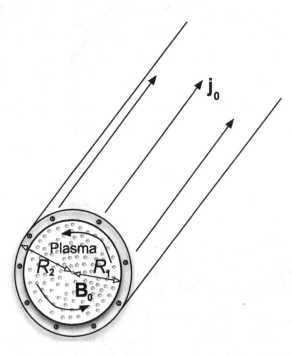

Figure 10.5 The z-pinch configuration: the plasma is confined in a cylinder by the action of an intense ring electric current $\mathbf{j}_0$ which circulates in a cylindrical envelope of thickness $R_2 - R_1$, and by a magnetic field induced by the current whose intensity is maximum in the envelope.

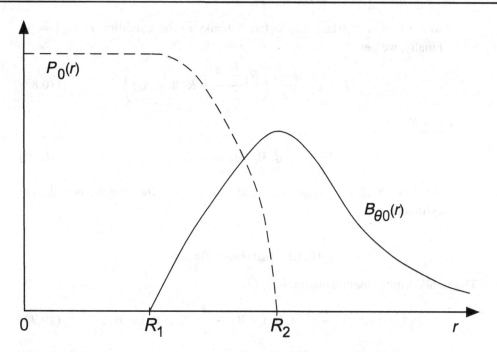

Figure 10.6 Schematic variation of the pressure and magnetic field in a *z*-pinch configuration at equilibrium.

equilibrium equation. We have demonstrated in Chapter 8 for this *z*-pinch configuration ($\forall r$) that

$$\frac{d}{dr}\left(P_0 + \frac{B_{0\theta}^2}{2\mu_0}\right) + \frac{B_{0\theta}^2}{\mu_0 r} = 0.$$ (10.84)

Let us solve this equilibrium equation (at the main order) in the special case where the current density is (not null and) constant only between R_1 and R_2 (see Figure 10.5). We have three regions (see Figure 10.6).

- $r < R_1$:

$$\oint_\Gamma \mathbf{B}_0 \cdot d\boldsymbol{\ell} = \iint_S (\nabla \times \mathbf{B}_0) \cdot d\mathbf{S} = \mu_0 \iint_S \mathbf{j}_0 \cdot d\mathbf{S} = 0,$$ (10.85)

and thus $B_{0\theta}(r) = 0$ and $P_0(r) = $ constant.
- $R_1 \le r \le R_2$:

$$\oint_\Gamma \mathbf{B}_0 \cdot d\boldsymbol{\ell} = \mu_0 j_0 \pi (r^2 - R_1^2),$$ (10.86)

and thus $B_{0\theta}(r) = \mu_0 j_0 (r^2 - R_1^2)/(2r)$. Returning to relation (10.84), one can deduce the following general expression for the pressure:

$$P_0(r) = -\frac{\mu_0 j_0^2}{2}\left(\frac{r^2}{2} - R_1^2 \ln(r)\right) + C,$$ (10.87)

where C is a constant that is fixed thanks to the condition $P_0(R_2) = 0$. Finally, we get

$$P_0(r) = \frac{\mu_0 j_0^2}{2} \left(\frac{R_2^2 - r^2}{2} + R_1^2 \ln(r/R_2) \right). \tag{10.88}$$

- $\underline{r > R_2}$:

$$\oint_\Gamma \mathbf{B}_0 \cdot d\boldsymbol{\ell} = \mu_0 I_0, \tag{10.89}$$

and thus $B_{0\theta}(r) = \mu_0 I_0 / 2\pi r$ and $P_0(r) = 0$ (no plasma outside the cylinder).

10.4.2   Instability Modes

The small-displacement equation is

$$\rho_0 \omega_k^2 \boldsymbol{\xi}_k = \nabla P_1 - \frac{1}{\mu_0}(\nabla \times \mathbf{B}_1) \times \mathbf{B}_0 - \frac{1}{\mu_0}(\nabla \times \mathbf{B}_0) \times \mathbf{B}_1, \tag{10.90}$$

with, from relations (9.45) and (9.48), respectively,

$$\mathbf{B}_1 = \nabla \times (\boldsymbol{\xi}_k \times \mathbf{B}_0),$$
$$P_1 = -\boldsymbol{\xi} \cdot \nabla P_0 - \gamma P_0 \nabla \cdot \boldsymbol{\xi}.$$

The imposed electric current circulates only between R_1 and R_2. It thus produces at the main order a magnetic confinement only for $r \geq R_1$. At first order, we obtain for the confined plasma ($r \leq R_1$)

$$-\rho_0 \omega_k^2 \boldsymbol{\xi}_k = -\nabla P_1 = \nabla(\boldsymbol{\xi} \cdot \nabla P_0 + \gamma P_0 \nabla \cdot \boldsymbol{\xi}). \tag{10.91}$$

We seek solutions in the form of a wave:

$$\boxed{\boldsymbol{\xi}_k = \boldsymbol{\xi}(r)\exp(i(m\theta + kz)).} \tag{10.92}$$

This form of solution is suggested by the symmetry of the problem (periodicity in θ and in z). The configuration analyzed leads to different instabilities which depend on the value of m. The most studied cases are

- $m = 0$, sausage instability;
- $m = 1$, kink instability;
- $m \geq 2$, fluting instability.

In the case where $m = 0$, we are in the configuration of Figure 10.7: local pinches appear by instability. For this case ($m = 0$), we have

$$\boldsymbol{\xi}_k(r, z) = [\xi_r(r)\mathbf{e_r} + \xi_\theta(r)\mathbf{e_\theta} + \xi_z(r)\mathbf{e_z}]\exp(ikz). \tag{10.93}$$

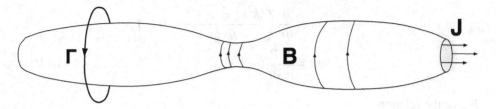

Figure 10.7 Result of the z-pinch instability for $m = 0$.

10.4.3   Resolution by Normal Modes (Case $m = 0$)

The small-displacement equation of the plasma inside the cylinder is written at first order $(r < R_1)$ as

$$-\rho_0\omega_k^2\boldsymbol{\xi}_k = \gamma P_0 \nabla(\nabla \cdot \boldsymbol{\xi}_k). \tag{10.94}$$

We recall that in this situation the pressure is constant and the confinement magnetic field does not penetrate into the plasma. This yields

$$-\rho_0\omega_k^2\xi_r \exp(ikz) = \gamma P_0 \partial_r(\nabla \cdot \boldsymbol{\xi}_k), \tag{10.95}$$

$$-\rho_0\omega_k^2\xi_\theta \exp(ikz) = 0, \tag{10.96}$$

$$-\rho_0\omega_k^2\xi_z \exp(ikz) = \gamma P_0 \partial_z(\nabla \cdot \boldsymbol{\xi}_k), \tag{10.97}$$

with

$$\nabla \cdot \boldsymbol{\xi}_k = \frac{1}{r}\frac{\partial}{\partial r}(r\xi_r)\exp(ikz) + \frac{\partial}{\partial z}(\xi_z \exp(ikz))$$

$$= \left(\frac{1}{r}\frac{\partial}{\partial r}(r\xi_r) + ik\xi_z\right)\exp(ikz). \tag{10.98}$$

We introduce the sound speed, $c_S^2 = \gamma P_0/\rho_0$, hence

$$-\omega_k^2\xi_r = c_S^2\partial_r\left(\frac{1}{r}\frac{\partial}{\partial r}(r\xi_r) + ik\xi_z\right), \tag{10.99}$$

$$\xi_\theta = 0, \tag{10.100}$$

$$-\omega_k^2\xi_z = c_S^2 ik\left(\frac{1}{r}\frac{\partial}{\partial r}(r\xi_r) + ik\xi_z\right). \tag{10.101}$$

The last relationship may be rewritten as

$$(k^2c_S^2 - \omega_k^2)\xi_z = \frac{ikc_S^2}{r}\frac{\partial}{\partial r}(r\xi_r). \tag{10.102}$$

The combination of Eqs. (10.99) and (10.102) gives

$$-\omega_k^2 \xi_r = c_S^2 \frac{\partial}{\partial r} \left(\frac{(k^2 c_S^2 - \omega_k^2)\xi_z}{ikc_S^2} + ik\xi_z \right)$$

$$= -\frac{1}{ik} \frac{\partial}{\partial r}(\omega_k^2 \xi_z),$$

(10.103)

hence the relation

$$\boxed{\frac{\partial \xi_z}{\partial r} = ik\xi_r \,.}$$

(10.104)

One finds

$$\frac{1}{r} \frac{\partial}{\partial r} \left(r \frac{\partial \xi_z}{\partial r} \right) = \frac{1}{r} \frac{\partial}{\partial r} (ikr\xi_r)$$

$$= \frac{ik}{r} \frac{\partial}{\partial r}(r\xi_r)$$

$$= \left(\frac{k^2 c_S^2 - \omega_k^2}{c_S^2} \right) \xi_z \,,$$

(10.105)

hence the equation

$$\boxed{\frac{\partial^2 \xi_z}{\partial r^2} + \frac{1}{r} \frac{\partial \xi_z}{\partial r} - A^2 \xi_z = 0 \,,}$$

(10.106)

with

$$A^2 = \frac{k^2 c_S^2 - \omega_k^2}{c_S^2} \,.$$

(10.107)

It is a differential equation of Bessel type whose solution is

$$\boxed{\xi_z(r) = I_0(Ar) \,,}$$

(10.108)

with I_0 the modified **Bessel function** of the first kind.

For $r < R_1$ we deduce that

$$\xi_r = \frac{1}{ik} \frac{\partial \xi_z}{\partial r} = \frac{A}{ik} I_1(Ar) \,.$$

(10.109)

The final resolution of the problem (derivation of the dispersion relation) is realized by taking into account the boundary conditions. We recall that ($\forall r$)

$$\frac{d}{dr} \left(P_0 + \frac{B_{0\theta}^2}{2\mu_0} \right) + \frac{B_{0\theta}^2}{\mu_0 r} = 0 \,.$$

(10.110)

Bessel function

The solution of the differential equation

$$\frac{d^2y(x)}{dx^2} + \frac{1}{x}\frac{dy(x)}{dx} - y(x) = 0 \qquad (10.111)$$

is the modified Bessel function of the first kind $y(x) = I_0(x)$, which is defined by

$$I_n(x) = \left(\frac{x}{2}\right)^n \sum_{p=0}^{\infty} \frac{(x/2)^{2p}}{2^{2p}p!\,(n+p)!}. \qquad (10.112)$$

In particular, one can notice the relations

$$\frac{dI_0}{dx} = I_1,$$

$$I_n(x) = i^{-n}J_n(ix),$$

where J_n is the Bessel function of the first kind. We give in Figure 10.8 the variation of some of Bessel functions. It may be noted that $I_0/I_1 > 1$.

We will consider the particular case where the electric current forms a cylindrical sheet and we integrate the equilibrium equation in the vicinity of this sheet at $r_0 = R_1 = R_2$:

$$\left[P_0 + \frac{B_{0\theta}^2}{2\mu_0}\right]_{r_0-\varepsilon}^{r_0+\varepsilon} = -\frac{1}{\mu_0}\int_{r_0-\varepsilon}^{r_0+\varepsilon} \frac{B_{0\theta}^2}{r}\,dr, \qquad (10.113)$$

with $P_0(r_0 + \varepsilon) = 0$ and $B_{0\theta}(r_0 - \varepsilon) = 0$, hence ($\varepsilon \to 0$)

$$\frac{B_{0\theta}^2(r_0^+)}{2\mu_0} - P_0(r_0^-) = -\frac{1}{\mu_0}\int_{r_0}^{r_0+\varepsilon} \frac{B_{0\theta}^2}{r}\,dr = 0, \qquad (10.114)$$

and therefore

$$\boxed{P_0(r_0^-) = \frac{B_{0\theta}^2(r_0^+)}{2\mu_0}}, \qquad (10.115)$$

which is the equilibrium relation for the pressure at r_0. Thus, in the vicinity of the interface we have

$$dP_0 = d\left(\frac{B_{0\theta}^2}{2\mu_0}\right) = \frac{\mathbf{B}_0}{\mu_0}\cdot d\mathbf{B}_0, \qquad (10.116)$$

We come back to the perturbation Eq. (9.31) for the pressure,

$$\frac{dP}{dt} + \gamma P \nabla \cdot \mathbf{u} = 0, \qquad (10.117)$$

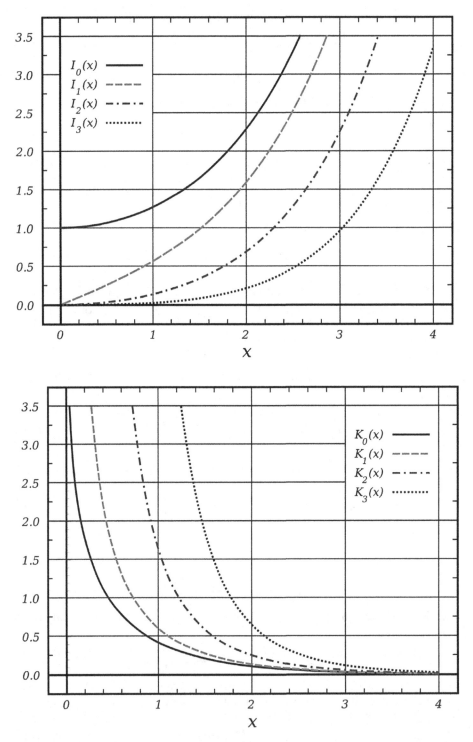

Figure 10.8 The first four entire modified Bessel functions of the first (I_n) and second (K_n) kinds.

that we integrate over a short time around the equilibrium:

$$P(\mathbf{r}, t) - P_0(\mathbf{r}_0, 0) = P(\mathbf{r}, t) - P_0 = -\gamma \int_0^t P \nabla \cdot \mathbf{u} \, dt$$

$$\simeq -\gamma P_0(\mathbf{r}_0, 0) \nabla \cdot \boldsymbol{\xi} \, . \qquad (10.118)$$

We recall that $\mathbf{r} = \mathbf{r}_0 + \boldsymbol{\xi}$, $\mathbf{u} = \dot{\boldsymbol{\xi}}$ and $\boldsymbol{\xi}(t = 0) = \mathbf{0}$. On the other hand, we have the relation

$$\frac{d\mathbf{B}}{dt} = \frac{\partial \mathbf{B}}{\partial t} + \mathbf{u} \cdot \nabla \mathbf{B} \, , \qquad (10.119)$$

which gives, after integration over a short time around the equilibrium and use of the relation (9.44),

$$\mathbf{B}(\mathbf{r}, t) - \mathbf{B}_0(\mathbf{r}_0, 0) = \int_0^t \frac{\partial \mathbf{B}}{\partial t} \, dt + \int_0^t \mathbf{u} \cdot \nabla \mathbf{B} \, dt$$

$$\simeq \int_0^t \dot{\boldsymbol{\xi}} \cdot \nabla \mathbf{B} \, dt$$

$$\simeq \boldsymbol{\xi} \cdot \nabla \mathbf{B}_0 \, . \qquad (10.120)$$

In this calculation, we have used the fact that the equilibrium magnetic field does not vary in time at the position $\mathbf{r}_0$. It yields

$$-\gamma P_0 \nabla \cdot \boldsymbol{\xi} = P(\mathbf{r}, t) - P_0$$

$$= \frac{\mathbf{B}_0}{\mu_0} \cdot (\mathbf{B}(\mathbf{r}, t) - \mathbf{B}_0)$$

$$\simeq \frac{\mathbf{B}_0}{\mu_0} \cdot (\boldsymbol{\xi} \cdot \nabla \mathbf{B}_0) \, . \qquad (10.121)$$

We recall that

$$-\rho_0 \omega_k^2 \xi_z \exp(ikz) = \gamma P_0 \, \partial_z (\nabla \cdot \boldsymbol{\xi}_k) \, , \qquad (10.122)$$

hence

$$\rho_0 \omega_k^2 \xi_z \exp(ikz) = \gamma P_0 \, \partial_z \left(\frac{\mathbf{B}_0}{\gamma P_0 \mu_0} \cdot (\boldsymbol{\xi} \cdot \nabla \mathbf{B}_0) \right)$$

$$= \partial_z \left(\frac{\mathbf{B}_0}{\mu_0} \cdot (\boldsymbol{\xi} \cdot \nabla \mathbf{B}_0) \right) \, , \qquad (10.123)$$

and thus

$$\frac{\rho_0 \omega_k^2}{ik} \xi_z \exp(ikz) = \frac{B_{0\theta}}{\mu_0} \left(\xi_r \exp(ikz) \frac{\partial}{\partial r} \left(\frac{\mu_0 I}{2\pi r} \right) \right)$$

$$\frac{\rho_0 \omega_k^2}{ik} \xi_z = -\frac{B_{0\theta}}{\mu_0} \left(\xi_r \frac{\mu_0 I}{2\pi r^2} \right)$$

$$\frac{\rho_0 \omega_k^2}{ik} \xi_z = -\frac{B_{0\theta}}{\mu_0} \left(\xi_r \frac{B_{0\theta}}{r} \right)$$

$$\omega_k^2 = -\frac{ik}{\mu_0 \rho_0 r} B_{0\theta}^2 \frac{\xi_r}{\xi_z}$$

$$\omega_k^2 = -\frac{ik}{\mu_0 \rho_0 r} B_{0\theta}^2 \frac{1}{ik} \frac{d\xi_z/dr}{\xi_z} \qquad (10.124)$$

$$\omega_k^2 = -\frac{B_{0\theta}^2}{\mu_0 \rho_0} \frac{1}{r\xi_z} \frac{d\xi_z}{dr}.$$

Finally, the solution of the problem is (with $r = r_0$)

$$\boxed{\omega_k^2 = -\frac{B_{0\theta}^2(r_0)}{\mu_0 \rho_0} \frac{A}{r_0} \frac{I_1(Ar_0)}{I_0(Ar_0)}}. \qquad (10.125)$$

From Figure 10.8, the ratio $I_0/I_1 > 1$ and therefore the term on the right-hand side of Eq. (10.125) can be **negative** (the coefficient A is positive for subsonic phase speeds, which is the case near the equilibrium). As a result, the z-pinch configuration is characterized by an **unstable** MHD equilibrium.

10.4.4   Configuration $m = 1$

We have seen that the complete resolution of the problem in the case $m = 0$ is tedious. The case $m = 1$ will therefore not be studied in detail; we will just give the geometric result. It can be shown that this always leads to an **unstable** equilibrium which deforms the cylinder as shown in Figure 10.9. This deformation is characterized by magnetic field lines closer near the concave side of the kink. This can easily be justified by the flux conservation (see Section 10.4.1).

10.5 z–θ Pinch Instability

We consider the z-pinch configuration in which we add a uniform component B_{0z} of the magnetic field in the cylinder. To do this, one needs an electric current in the z direction but also in the θ direction; this is called the z–θ pinch. In this case, the plasma can be **stabilized**.

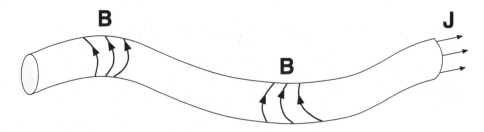

Figure 10.9 Result of the kink instability for $m = 1$.

As for the z-pinch configuration, we seek general solutions like

$$\xi_k = \xi(r)\exp(i(m\theta + kz)),$$

(10.126)

whose form is suggested by the symmetry of the problem. As in the previous case, the plasma presents different instabilities depending on the value of m. We will be interested only in the case $m = 0$, for which we do not redo the calculations. It can be shown that the dispersion relation is

$$\omega_k^2 = \frac{k^2 B_{0z}^2}{\mu_0\rho_0} - \frac{B_{0\theta}^2(r_0)}{\mu_0\rho_0}\frac{\tilde{A}}{r_0}\frac{I_1(\tilde{A}r_0)}{I_0(\tilde{A}r_0)},$$

(10.127)

with

$$\tilde{A} = k^2\left(1 + \frac{(\omega/k)^4}{c_S^2 b_0^2 - (\omega/k)^2(c_S^2 + b_0^2)}\right)$$

(10.128)

and b_0 the Alfvén speed. We see that if $B_{0z} \to 0$ the Alfvén speed becomes zero and $\tilde{A} \to A$: we recover the z-pinch situation. The second effect of the uniform magnetic field in the cylinder is to provide another term in expression (10.127) that is **positive**. This term is therefore **stabilizing**. Figure 10.10 illustrates this situation of a z–θ pinch.

10.6 Magneto-rotational Instability in Accretion Disks

In this section, we shall study the magneto-rotational instability, which is particularly important for accretion disks around a massive and compact central object such as a star in formation, a neutron star, or a black hole.

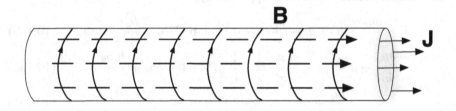

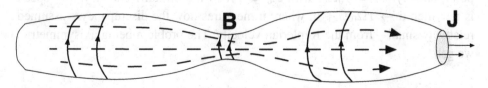

Figure 10.10 The z–θ pinch configuration for $m = 0$. The axial component of the magnetic field is shown as dashed lines.

An accretion disk is composed of plasma, gas, and dust in orbit around a central object (see Figure 10.11). As a first approximation, the material of the disk is in equilibrium between the gravity of the central object (neglecting the self-gravity of the disk) and the centrifugal force. The material then follows Kepler's third law, which, for a circular rotation, gives a Keplerian velocity that varies as $r^{-1/2}$. Insofar as the material moves *a priori* in a viscous medium, the radial layers of the disk rub against one another and convert mechanical energy into heat. The disk material then falls on to the central object, emitting radiation due to heating: this is basically a mechanism of conversion of gravitational energy into radiative energy. For example, in the case of active galaxy nuclei (AGN), the light power can reach up to 10^{42} J/s. In most cases, the accretion is accompanied by a mass ejection in the form of jets perpendicular to the plane of the disk (see Figure 10.11). In order to break the balance between the centrifugal force and gravity to accrete material to the central object, it is necessary to introduce a physical process capable of extracting angular momentum from rotating matter. As mentioned above, the viscous friction can contribute, but some estimates show in fact that the viscosity is too weak – by several orders of magnitude – to explain the observed phenomena. Among other possible processes turbulence appears to be an effective way to transport the material. The question of its generation was a problem until recently because of the absence of relevant linear hydrodynamic instability. It was in 1991 (Balbus and Hawley, 1991) that the magneto-rotational instability was re-discovered[1] and proposed for magnetized accretion disks as the primary source of turbulence and of transport of matter.

We will demonstrate the existence of the magneto-rotational instability in the simplified framework of the (non-relativistic) standard, incompressible, ideal and inviscid MHD. We will consider a thin accretion disk; in cylindrical coordinates we have for the velocity field (see Appendix 2)

$$\frac{\partial u_r}{\partial t} + \mathbf{u} \cdot \nabla u_r - \frac{u_\theta^2}{r} = -\partial_r P_* + \mathbf{b} \cdot \nabla b_r - \frac{b_\theta^2}{r} - \partial_r \Phi, \quad (10.129)$$

$$\frac{\partial u_\theta}{\partial t} + \mathbf{u} \cdot \nabla u_\theta + \frac{u_r u_\theta}{r} = -\frac{1}{r}\partial_\theta P_* + \mathbf{b} \cdot \nabla b_\theta + \frac{b_r b_\theta}{r}, \quad (10.130)$$

$$\frac{\partial u_z}{\partial t} + \mathbf{u} \cdot \nabla u_z = -\partial_z P_* + \mathbf{b} \cdot \nabla b_z. \quad (10.131)$$

We will make a local study around a reference point of the disk located at $r = r_0$ (see Figure 10.12) and we will take as a reference speed the rotation speed at that point, namely $u_{0\theta} = r_0 \Omega_0$, with $\Omega_0 = \Omega(r_0)$ the orbital frequency. The speed is rewritten $\mathbf{u} \to r\Omega_0 \mathbf{e}_\theta + \mathbf{u}$, where $\mathbf{u}$ measures now the discrepancy – assumed relatively small – from the Keplerian velocity. The problem being axisymmetric, we get

---

[1] Actually E. Velikhov (1959) and S. Chandrasekhar (1960) had been the first to highlight the existence of the magneto-rotational instability.

Figure 10.11 The observation (top) of the elliptical galaxy Centaurus A in the visible, microwave, and X-ray domains reveals the presence of extra-galactic jets ending in lobes. (Credit: ESO/WFI (visible); MPIfR/ESO/APEX/A. Weiss *et al.* (microwave); NASA/CXC/CfA/R. Kraft *et al.* (X-ray).) The theoretical modeling of these jets is generally done in the framework of MHD (Ferreira, 1997). A zoom (bottom) around the center of the galaxy shows the position of a super-massive black hole – of mass greater than 10^7 times the mass of the Sun – at the origin of the jets. (Credit: NASA/TANAMI/Müller *et al.*)

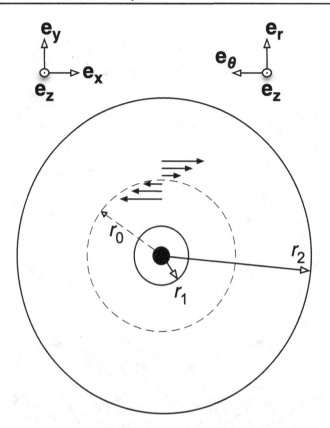

Figure 10.12 An accretion disk between r_1 and r_2 around a solid and compact object (a black hole). r_1 is supposed large enough for us not to take into account relativistic effects due to the central object. The effect of the gravitational tide is equivalent to a shear in the rotating frame of reference.

$$\frac{\partial u_r}{\partial t} + \mathbf{u} \cdot \nabla u_r - \frac{u_\theta^2}{r} - 2\Omega_0 u_\theta - r\Omega_0^2 = -\partial_r P_* + \mathbf{b} \cdot \nabla b_r - \frac{b_\theta^2}{r} - \partial_r \Phi , \quad (10.132)$$

$$\frac{\partial u_\theta}{\partial t} + \mathbf{u} \cdot \nabla u_\theta + \frac{u_r u_\theta}{r} + 2\Omega_0 u_r = -\frac{1}{r} \partial_\theta P_* + \mathbf{b} \cdot \nabla b_\theta + \frac{b_r b_\theta}{r} , \quad (10.133)$$

$$\frac{\partial u_z}{\partial t} + \mathbf{u} \cdot \nabla u_z = -\partial_z P_* + \mathbf{b} \cdot \nabla b_z , \quad (10.134)$$

or

$$\frac{\partial \mathbf{u}}{\partial t} + \mathbf{u} \cdot \nabla \mathbf{u} + 2\Omega_0 \mathbf{e_z} \times \mathbf{u} = -\nabla P_* + \mathbf{b} \cdot \nabla \mathbf{b} + (r\Omega_0^2 - \partial_r \Phi)\mathbf{e_r} , \quad (10.135)$$

where we recognize the Coriolis force, $-2\Omega_0 \mathbf{e_z} \times \mathbf{u}$, and the tide term, $(r\Omega_0^2 - \partial_r \Phi)\mathbf{e_r}$. At equilibrium it is assumed that the radial gravitational force and the centrifugal force dominate, and therefore at any point on the disk ($r_1 \leq r \leq r_2$)

$$r\Omega^2 = \partial_r\Phi\,, \tag{10.136}$$

where $\Phi = -\mathcal{G}M/r$, $\mathcal{G}$ is the gravitational constant, and M is the mass of the central object. In particular, around the reference point we have at first order

$$r\Omega_0^2 - \partial_r\Phi = r(\Omega_0^2 - \Omega^2) = -2\Omega_0 r_0(r - r_0)\partial_r\Omega|_{r_0} = 2\Omega_0(r - r_0)S_0\,, \tag{10.137}$$

where $S_0 = -r_0\,\partial_r\Omega|_{r_0}$ is interpreted as the mean shear of the laminar flow due to the effect of the gravitational tide. Finally, one introduces the local Cartesian coordinates centered on the point of reference in r_0 (see Figure 10.12). This results in the system

$$\frac{\partial\mathbf{u}}{\partial t} + \mathbf{u}\cdot\nabla\mathbf{u} + 2\Omega_0\mathbf{e_z}\times\mathbf{u} = -\nabla P_* + \mathbf{b}\cdot\nabla\mathbf{b} + 2\Omega_0 yS_0\mathbf{e_y}\,, \tag{10.138}$$

$$\frac{\partial\mathbf{b}}{\partial t} + \mathbf{u}\cdot\nabla\mathbf{b} = \mathbf{b}\cdot\nabla\mathbf{u}\,, \tag{10.139}$$

where the induction equation remains unchanged from previous handling.

We seek the linear solutions of the system (10.138)–(10.139). To do this, it is useful to note that $\mathbf{u} = S_0 y\,\mathbf{e_x}$ is a particular solution of the linear equations. The linear solutions $(\mathbf{u}_1, \mathbf{b}_1)$ will be obtained around this shear solution in the case where the accretion disk is crossed by a uniform magnetic field $b_0\mathbf{e_z}$. The linearized axisymmetric system can be written as

$$\frac{\partial\mathbf{u}_1}{\partial t} = (2\Omega_0 - S_0)u_{1y}\mathbf{e_x} - 2\Omega_0 u_{1x}\mathbf{e_y} - \nabla P_* + b_0\,\partial_z\mathbf{b}_1\,, \tag{10.140}$$

$$\frac{\partial\mathbf{b}_1}{\partial t} = b_0\,\partial_z\mathbf{u}_1 + S_0 b_{1y}\mathbf{e_x}\,. \tag{10.141}$$

One introduces the notation

$$\mathbf{u}_1 \rightarrow \mathbf{u}_1\exp(i(k_y y + k_z z - \omega t))\,, \tag{10.142}$$

$$\mathbf{b}_1 \rightarrow \mathbf{b}_1\exp(i(k_y y + k_z z - \omega t))\,, \tag{10.143}$$

hence

$$-i\omega\mathbf{u}_1 = (2\Omega_0 - S_0)u_{1y}\mathbf{e_x} - 2\Omega_0 u_{1x}\mathbf{e_y} - ik P_* + ik_z b_0\mathbf{b}_1\,, \tag{10.144}$$

$$-i\omega\mathbf{b}_1 = +ik_z b_0\mathbf{u}_1 + S_0 b_{1y}\mathbf{e_x}\,, \tag{10.145}$$

and finally

$$(\omega^2 - k_z^2 b_0^2)\mathbf{b}_1 = i\omega 2\Omega_0 b_{1y}\mathbf{e_x} - k_z b_0 P_*\mathbf{k} - 2\Omega_0(S_0 b_{1y} + i\omega b_{1x})\mathbf{e_y}\,. \tag{10.146}$$

After some manipulation, we obtain the dispersion relation

$$\omega^4 - (\alpha^2\kappa^2 + 2k_z^2 b_0^2)\omega^2 - k_z^2 b_0^2(2\Omega_0 S_0\alpha^2 - k_z^2 b_0^2) = 0\,, \tag{10.147}$$

where by definition $\kappa^2 \equiv 2\Omega_0(2\Omega_0 - S_0)$ is the epicyclic frequency and $\alpha \equiv k_z/k$. The general solution takes the form

$$\omega_\pm^2 = k_z^2 b_0^2 + \frac{\alpha^2 \kappa^2}{2} \pm \frac{1}{2}\sqrt{\alpha^4 \kappa^4 + 16 k_z^2 b_0^2 \Omega_0^2 \alpha^2} \,. \tag{10.148}$$

The study of the solution ω_-^2 reveals the presence of an unstability under certain conditions. For example, when the magnetic field is relatively strong, the stability condition $\omega_-^2 > 0$ leads to the relationship

$$k b_0 > \sqrt{2\Omega_0 S_0} = \sqrt{\frac{4}{3} S_0} \,, \tag{10.149}$$

the last equality being valid for a Keplerian disk (for which $S_0 = 3\Omega_0/2$ and thus $\kappa^2 = 4S_0^2/9$). Therefore, an intense magnetic field b_0 stabilizes the disk. A relatively weak field b_0 allows, however, the emergence of an instability – **the magneto-rotational instability** – which is confined to the largest scales, i.e. to small k. Therefore, there is a range of values of b_0 for which the accretion disk is unstable. Finally, it can be shown easily that in the Keplerian regime the growth rate of the magneto-rotational instability is maximum when

$$k b_0 = \sqrt{\frac{5}{12} S_0} \,. \tag{10.150}$$

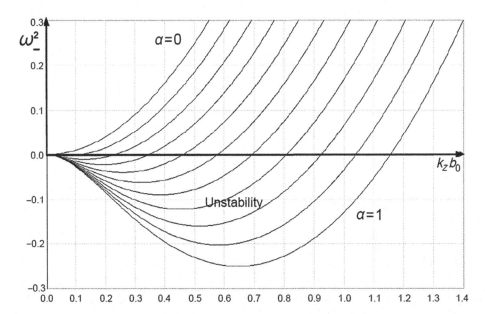

Figure 10.13 Solution ω_-^2 for $S_0 = 1$ depending on the parameter α (ranging from 0 to 1 with constant increments of 0.1). The magneto-rotational instability disappears systematically for strong fields.

In the particular case where $\alpha = 1$, one gets $\omega_-^2 = -S_0^2/4$, hence the growth rate is

$$\gamma_{max} = S_0/2. \qquad (10.151)$$

This value is relatively high: to realize that, one has to evaluate the instability amplification over one period of rotation of the disk, i.e. $\tau = 2\pi r_0/u_{0\theta} = 2\pi/\Omega_0$. This yields

$$\exp(\gamma_{max}\tau) = \exp\left(\frac{3\pi}{2}\right) > 110. \qquad (10.152)$$

On this time scale, the magneto-rotational instability is explosive. The solution ω_-^2 for $S_0 = 1$ is plotted in Figure 10.13 for different values of α. One notes that under these conditions the magneto-rotational instability disappears systematically for strong fields.

Exercise for Part III

III.a Resistive Tearing Instability

We propose to study a resistive instability called tearing instability within the framework of the standard, incompressible and inviscid MHD. In this problem, we consider the equilibrium configuration of Figure III.a.1 in which a magnetic field $\mathbf{b}_0(y)$ changes its direction on a length scale L. A thin resistive layer of thickness ϵL (with $\epsilon \ll 1$) is present around the resonant surface defined by $\mathbf{k} \cdot \mathbf{b}_0 = 0$.

(1) In this problem, we seek perturbative solutions of the form

$$\mathbf{X} = \mathbf{X}(y) \exp[i(kz - \omega t)].$$

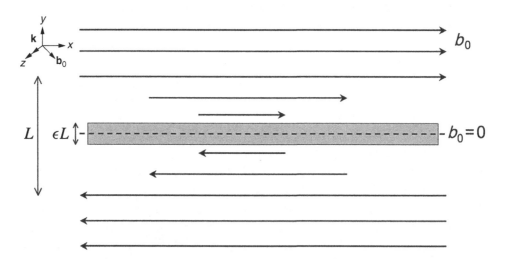

Figure III.a.1 The equilibrium configuration of a magnetic field reversal $\mathbf{b}_0$ on a length scale L. The shaded part represents the resistive layer of thickness ϵL.

Demonstrate that the associated small-displacement equation can be written

$$\omega^2[\xi_y'' - k^2\xi_y] = -i(\mathbf{k} \cdot \mathbf{b}_0)\left[b_{1y}'' - k^2 b_{1y} - b_{1y}\frac{(\mathbf{k} \cdot \mathbf{b}_0)''}{(\mathbf{k} \cdot \mathbf{b}_0)}\right],$$

where the first and second derivatives in y are denoted by single and double primes, respectively.

(2) Demonstrate that the associated magnetic field equation is

$$b_{1y} = i(\mathbf{k} \cdot \mathbf{b}_0)\xi_y + \frac{i\eta}{\omega}(b_{1y}'' - k^2 b_{1y}).$$

(3) Insofar as the resistive term of the previous equation can be assumed significant only within the resistive layer (η is assumed to be small), what is the expression for ω away from the resistive layer?

(4) We will now look at the solution associated with the resistive layer. We have to solve a fourth-order equation. To simplify the analysis, we will assume that the dominant terms inside the resistive layer are those associated with the perturbation terms of the higher-order derivatives. Justify this assumption and then demonstrate that

$$\omega \simeq +i\frac{(\mathbf{k} \cdot \mathbf{b}_0)^2(\epsilon L)^2}{\eta}.$$

Subsequently we will write $\omega = iA$, with $A > 0$.

(5) We will evaluate the thickness ϵL of the resistive layer from the relation obtained in question (2) balancing the convective and resistive terms. Demonstrate that

$$\epsilon L \simeq \left(\frac{\eta A}{k^2(b_0'(0))^2}\right)^{1/4},$$

where we have used the relation $\mathbf{k} \cdot \mathbf{b}_0(\epsilon L) \simeq \mathbf{k} \cdot \mathbf{b}_0(0) + \epsilon L \mathbf{k} \cdot \mathbf{b}_0'(0) \simeq \epsilon L k b_0'(0)$.

(6) We want to evaluate the time scale of variations of the tearing instability. To do this, we introduce the Alfvén time $\tau_A = L/b_0$, the resistive time $\tau_R = L^2/\eta$, and the mean relation

$$b_{1y}'' \simeq \frac{b_{1y}'(\epsilon L/2) - b_{1y}'(-\epsilon L/2)}{\epsilon L} \simeq \frac{K}{\epsilon L}b_{1y}(0).$$

Show that

$$A\tau_A \simeq (LK)^{4/5}(Lk)^{2/5}(\tau_A/\tau_R)^{3/5}.$$

Part IV

Turbulence

11

Hydrodynamic Turbulence

Turbulence is generally associated with the formation of vortices in a fluid. There are numerous experiences in daily life where one can note the presence of turbulence: the movements of a river downstream of an obstacle, the smoke escaping through a chimney, vortical motions of the air, or the turbulence zones that we sometimes cross by plane. Since it is not necessary to use powerful microscopes or telescopes to study turbulence one could conclude that it is probably not difficult to understand it. Unfortunately that is not the case! Although significant progress has been made since the middle of the twentieth century, several important questions remain unanswered and it is clear that at the beginning of the twenty-first century turbulence remains a central research topic in physics.

The first theoretical bricks of turbulence were laid from the moment physicists started to tackle the non-linearities of the hydrodynamic equations. As we will see, it is in this context that the first fundamental law of turbulence was established: this was the statistical law of Kolmogorov (1941). Nowadays the theoretical treatment of turbulence is partly based on numerical simulations which, accompanied by very powerful tools of visualization, allow us to tackle this difficult problem from a different angle and stimulate new questions. The purpose of this chapter is to present concepts and fundamental results on fully developed turbulence. This chapter is devoted to hydrodynamics, from which some foundations of the theory of turbulence have emerged. The two other chapters in this part of the book will be devoted to MHD turbulence.

11.1 What is Turbulence?

11.1.1 Unpredictability and Turbulence

It is not easy to define turbulence quantitatively because to do this one requires knowledge of a number of concepts that will be defined partly in this chapter.

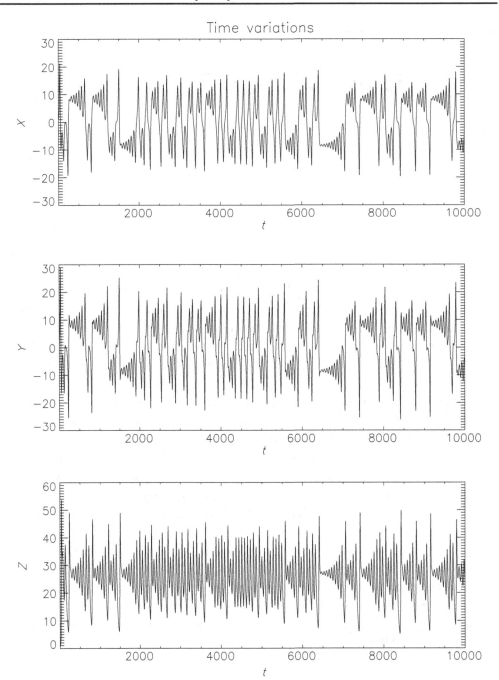

Figure 11.1 Simulation of the Lorenz system with $\sigma = 10$, $\rho = 28$, and $\beta = 8/3$.

Without going into the details, we can notice that the disordered – or chaotic – aspect seems to be the main characteristic of turbulent flows. It is often said that a system is chaotic when two points originally very close to each other in phase space separate exponentially over time. As we will see later, this definition can be extended to the case of fluids.

The media success of chaos theory dates back to the year 1961. This was the year in which the meteorologist Edward Lorenz from MIT decided to use his computer (a Royal McBee LGP-300, without a screen) capable of performing 60 calculation operations per second to numerically integrate a very simplified system of non-linear differential equations – the Lorenz system (see Figure 11.1): this is a simplified version of the fluid equations for thermal convection. He observed[1] by chance that two very similar initial conditions diverge pretty quickly (Lorenz, 1963). Since the linear functions involved results proportional to the initial uncertainties, the observed difference could only be explained by the presence of the non-linear terms in the model equations. Then, Lorenz understood that, even though some non-linear phenomena are governed by rigorous and perfectly deterministic equations, accurate predictions are impossible because of the sensitivity to the initial conditions, which is (as we know) a major problem in meteorology. To make this result clear, Lorenz used (in 1972) an image that contributed to the media success of the chaos theory: that of his famous butterfly effect. He explained that the laws of meteorology are so sensitive to initial conditions that the simple beat of butterfly wings in Brazil could trigger a tornado in Texas, which means in other words that the future is **unpredictable**. But what is unpredictable is not necessarily chaotic (i.e. disordered), as we can see from the existence of strange attractors in deterministic chaos. In the phase space, this is illustrated by trajectories that are irresistibly attracted by complex geometric figures. These systems wander randomly around these figures without returning a second time to the same point (see Figure 11.2). Note that a similar system of ordinary differential equations can be used in the context of the dynamo problem in order to investigate the origin of the Earth's magnetic reversals over time scales of the order of millions of years (see Chapter 4 and Cook and Roberts (1970)).

The Lorenz system

In 1963 (Lorenz, 1963), Lorenz proposed the following system with three degrees of freedom:

$$\frac{dX}{dt} = \sigma(Y - X), \tag{11.1}$$

$$\frac{dY}{dt} = \rho X - Y - XZ, \tag{11.2}$$

$$\frac{dZ}{dt} = XY - \beta Z, \tag{11.3}$$

where σ, ρ, and β are real parameters. In Figure 11.2, we see that this dynamical system presents a strange attractor.

---

[1] Edward Lorenz was not the first to wonder about unpredictability: Henri Poincaré addressed this issue at the end of the nineteenth century in his stability study of the solar system (Poincaré, 1889).

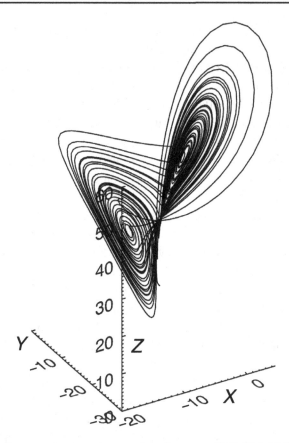

Figure 11.2 The Lorenz attractor appears when the curve $(X(t), Y(t), Z(t))$ is plotted (same data as in Figure 11.1).

Turbulent flows do not escape the unpredictability: two very similar initial conditions diverge pretty quickly over time. Although the equations governing the fluid motions are deterministic (like in MHD), it is not possible to predict exactly the state of the fluid at a much later moment. A distinction exists, however, between turbulence and chaos: chaos is nowadays mainly used in mechanics to describe a deterministic dynamical system with a limited (small) number of degrees of freedom. In turbulence, the flows have a very large number of degrees of freedom, which means in general that a very large number of spatial scales will be excited. Nevertheless, turbulence is predictable in the statistical sense, hence the importance of studying turbulence with statistical tools.

11.1.2   Transition to Turbulence

The observation of turbulence in fluid mechanics is often the common experience in real life. Indeed, it is under this regime that most usual natural terrestrial fluids such as air and water are found. There exists a wide variety of turbulent flows: geophysical flows (atmospheric wind, river currents), astrophysical flows (gas circulation around planets, solar photospheric plasma, solar and stellar winds, and interstellar molecular clouds), tokamak plasmas, biological (blood) and quantum (superfluid) fluids, or even industrial flows (aerospace, hydraulic, chemical). Despite this diversity, these turbulent flows have a number of common properties.

The most familiar example of turbulent flow is probably that of a watercourse encountering an obstacle like a rock. We then observe downstream a random water movement characterized by the presence of eddies of different sizes. As we will see, the eddy is a central concept in the analysis of turbulence and, in particular, in the definition of an energy cascade towards scales that are generally smaller. In Figure 11.3, we see schematically how such a flow goes from the laminar low-Reynolds-number regime to the fully developed turbulence regime for which the Reynolds number exceeds 1000. In particular, during this transition the so-called Kármán vortex street is formed at $R_e \sim 100$. Historically, it was O. Reynolds who was in 1883 the first to study the transition between these two regimes, hence the name given to the dimensionless parameter – **the Reynolds number** – measuring the degree of turbulence in a flow. This number reflects the relative importance of non-linear effects compared with dissipative effects. It can be written as

$$R_e = \frac{UL}{\nu},$$
(11.4)

where U and L are the characteristic velocity and length of the fluid. The case of a plasma was already illustrated in Figure 6.3, where a measure of the magnetic field in different regions of the magnetosphere is reported. Clearly, we see that the magnetosheath is turbulent, with magnetic fluctuations of different amplitudes, while a regime close to laminar is observed in the magnetopause.

11.2 Statistical Tools and Symmetries

We have discussed the importance of addressing the problem of turbulence by statistical tools to gain a better understanding of its randomness. In this Section, we recall some of these tools that are introduced generally in a course on statistical physics.

11.2.1 Ensemble Average

The ensemble average $\langle X \rangle$ of a quantity X is a statistical average performed on N independent realizations (with $N \to +\infty$) where we measure this quantity:

$$\langle X \rangle = \lim_{N \to +\infty} \frac{1}{N} \sum_{n=1}^{N} X_n.$$
(11.5)

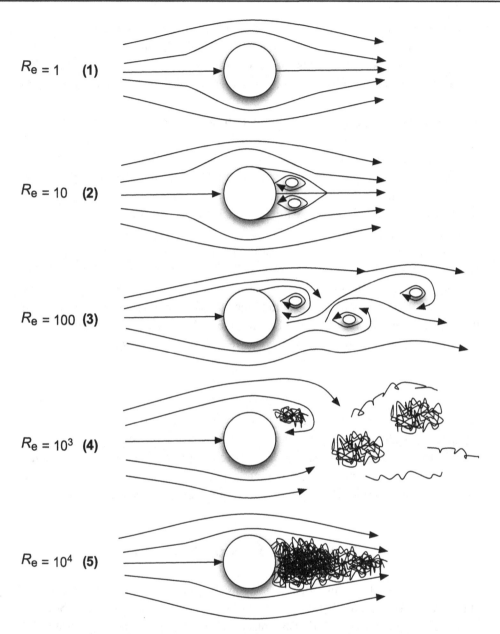

$R_e = 1$ **(1)**

$R_e = 10$ **(2)**

$R_e = 100$ **(3)**

$R_e = 10^3$ **(4)**

$R_e = 10^4$ **(5)**

Figure 11.3 Transition between the laminar and turbulent regimes (measured with the Reynolds number R_e) for a flow coming from the left and bumping into a cylindrical obstacle. Adapted from Feynman *et al.* (1964).

If the averaged quantity is, for example, the velocity field, one has

$$\langle \mathbf{u}(\mathbf{x}, t) \rangle = \lim_{N \to +\infty} \frac{1}{N} \sum_{n=1}^{N} \mathbf{u}_n(\mathbf{x}, t). \tag{11.6}$$

The averaging operation commutes with derivatives of different kinds, for example

$$\left\langle \frac{\partial \mathbf{u}(\mathbf{x}, t)}{\partial \mathbf{x}} \right\rangle = \frac{\partial \langle \mathbf{u}(\mathbf{x}, t) \rangle}{\partial \mathbf{x}}. \tag{11.7}$$

This ensemble average operator is analogous to the one used in statistical thermodynamics. Generally, this is not equivalent to a temporal or a spatial average, except under certain conditions. For example, when turbulence is homogeneous the ergodic hypothesis allows one to calculate an ensemble average as a spatial average (Galanti and Tsinober, 2004). Note that no proof of the ergodic theorem is known for the Navier–Stokes or MHD equations.

11.2.2 Autocorrelation

To characterize the disorder in a signal $u(x, t)$ one uses the concept of correlation. The simplest correlation function is the autocorrelation:[2]

$$R(x, t, T) = \langle u(x, t) u(x, t + T) \rangle, \tag{11.8}$$

which measures the similarity of a function with itself, here at two different moments. The quantity $u(x, t)$ (e.g. a component of the velocity) is a random function. To get statistical independence between $u(x, t)$ and $u(x, t + T)$, T cannot be too small because it is expected *a priori* that the signal has some memory: then, T must be larger than a value T_c called the **correlation time**. A similar analysis can be done for two measures not in time but in space. In this case, we arrive at the concept of the **correlation length** L_c (also called the integral scale). Thus, a random flow is characterized by a spatio-temporal memory whose horizon is measured by T_c and L_c. The study of turbulence consists of extracting information about the spatio-temporal memory of the flow which cannot be revealed if we are working on spatial and temporal correlation scales that are relatively too large. Figure 11.4 shows the concept of the correlation time; by definition, we have

$$T_c = \frac{1}{R(0)} \int_0^{+\infty} R(T) dT, \tag{11.9}$$

where we can forget the dependence on t (or x) by virtue of the assumption of homogeneity.

11.2.3 Probability Distribution and PDF

Let us define $F_y(x)$ the probability of finding a fluctuation of the random variable y in the interval $]-\infty, x]$: the function F_y is by definition a probability distribution. From this definition one has

- $F_y(x)$ is an increasing function,
- $F_y(x)$ is a continuous function,
- $F_y(-\infty) = 0$ and $F_y(+\infty) = 1$.

---

[2] It was the British scientist Francis Galton (1822–1911) who seems to have been the first to properly introduce the concept of correlation for statistical studies in biology.

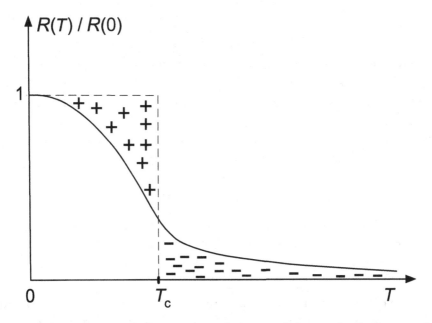

Figure 11.4 Illustration of the meaning of the correlation scale T_c from a given autocorrelation function: the surface $+$ is by definition equal to the surface $-$ (see relation (11.9)).

If this function is differentiable then $F'_y(x)$ defines a probability density function (PDF), i.e. $F'_y(x)$ is the probability of finding y in the interval $]x, x + dx]$. In the framework of intermittency we will see that the normal (or Gaussian) and Poisson PDF play a central role.

11.2.4   Moments and Cumulants

The moments of a probability density function are the means of the powers:

$$M_n = \langle y^n \rangle = \int_{-\infty}^{+\infty} x^n F'_y(x) dx . \tag{11.10}$$

We may note that the moment of order one is the mean or expected value. The moments of $y - \langle y \rangle$ are said to be centered. The variance is the centered moment of order two:

$$\langle (y - \langle y \rangle)^2 \rangle , \tag{11.11}$$

while the mean quadratic deviation is the root of the variance.

Given any non-Gaussian random function of zero mean whose second-order moments are known, it is then possible to calculate the fictitious moments of order n that this function would have if it were a Gaussian function. The difference between the actual nth-order moment of the function and the corresponding Gaussian value is called the nth-order cumulant. Then, the odd cumulants are equal to the moments (since the odd moments of a Gaussian are zero) and by definition all the cumulants are zero for a Gaussian function.

11.2.5 Structure Functions

A structure function of order n of a quantity $f(\mathbf{x})$ is by definition

$$S_n = \langle (f(\mathbf{x}_1) - f(\mathbf{x}_2))^n \rangle = \langle (\delta f)^n \rangle, \qquad (11.12)$$

where $\mathbf{x}_1$ and $\mathbf{x}_2$ are two points of the space. We will see below that the first rigorous law established in turbulence by Kolmogorov (1941) involves the velocity structure function of order three.

11.2.6 Symmetries

In order to simplify the analytical study of turbulence, we often impose certain symmetries on the flow. These symmetries are taken in the statistical sense.

- **Homogeneity**: this is the space translation invariance. This is the most classical assumption that is satisfied at the heart of turbulence, i.e. far from the walls of an experiment. In astrophysics it is often a good approximation. This assumption is essential in the theoretical treatment of turbulence insofar as it brings important simplifications both in physical space and in Fourier space. For a homogeneity turbulence the ergodic hypothesis allows one to calculate an ensemble average as a spatial average.

- **Stationarity**: this is the time translation invariance. This is a very classical hypothesis insofar as a system generally finds its balance between the external forces and the dissipation which occurs at a small–scale by viscous friction. For stationary turbulence the ergodic hypothesis allows one to calculate an ensemble average as a time average.

- **Isotropy**: this is the statistical invariance under any arbitrary rotation. This is a classic assumption in hydrodynamics that is less justified in MHD.

- **Mirror symmetry**: this is the statistical invariance under any plane symmetry. This corresponds to an invariance when the sign of all vectors ($\mathbf{x} \to -\mathbf{x}$, $\mathbf{u} \to -\mathbf{u}$, etc.) is changed. It allows the removal of quantities such as the kinetic helicity. One speaks of strong isotropy when turbulence is both isotropic and mirror symmetric. Below, we shall use the word **isotropy** in the weak sense to indicate invariance under rotations but not necessarily under reflections of the frame of reference.

- **Scale-invariance**: this is the (non-statistical) invariance by a transformation of the type $\mathbf{u}(\mathbf{x}, t) \rightarrow \lambda^h \mathbf{u}(\lambda \mathbf{x}, \lambda^{1-h} t)$. The solutions of the Navier–Stokes equations satisfy this symmetry if $h = -1$. If the viscosity is zero then h can be anything. In practice, this symmetry can be found in the turbulent regime if the scales considered are much greater than those at which the viscosity acts.

11.3   The Exact laws of Kolmogorov

The first exact law of fully developed turbulence was obtained by Kolmogorov (1941). This is called the four-fifths law, simply because of the value of the constant that appears. The path taken by Kolmogorov requires the use of tensor analysis which – although elegant – tends to weigh down the formalism. In this section, we choose to follow a more direct path which allows us to derive this law in a different form: this is the four-thirds law, which was obtained only in 1997 (Antonia *et al.*, 1997). These laws express how third-order structure functions for the velocity are related to the distance between the two points of measurement in the case of a homogeneous, isotropic, and three-dimensional incompressible turbulence.

11.3.1   The Kármán–Howarth Equations

Let us write the Navier–Stokes equations at points $\mathbf{x}$ and $\mathbf{x}'$ for the components i and j, respectively (by virtue of the incompressibility, $\partial_k u_k = 0$):

$$\partial_t u_i + \partial_k (u_k u_i) = -\partial_i P + \nu \, \partial_{kk}^2 u_i + f_i, \tag{11.13}$$

$$\partial_t u_j' + \partial_k' (u_k' u_j') = -\partial_j' P' + \nu \, \partial_{kk}'^2 u_j' + f_j', \tag{11.14}$$

where Einstein's notations are used. To simplify the writing of equations, we define $\mathbf{u}(\mathbf{x}) \equiv \mathbf{u}$, $\mathbf{u}(\mathbf{x}') \equiv \mathbf{u}'$, and $\partial/\partial \mathbf{x}' \equiv \partial'$, and we will consider only velocity fluctuations with zero average, i.e. $\langle \mathbf{u} \rangle = 0$. Subsequently, we will forget for a while the external force $\mathbf{f}$. Multiply the first equation by u_j' and the second by u_i. The addition of these two equations gives us, on taking the ensemble average, a dynamic equation for the second-order correlation tensor:

$$\partial_t \langle u_i u_j' \rangle + \langle \partial_k (u_k u_i u_j') + \partial_k' (u_k' u_j' u_i) \rangle = -\langle \partial_i (P u_j') + \partial_j' (P' u_i) \rangle$$
$$+ \nu \langle \partial_{kk}^2 (u_i u_j') + \partial_{kk}'^2 (u_j' u_i) \rangle. \tag{11.15}$$

To obtain this result, we have used the relation: $\partial_k u_j' = \partial_k' u_i = 0$. Insofar as turbulence is homogeneous, the two-point correlation tensors depend only on the relative distance $\boldsymbol{\ell}$, where $\mathbf{x}' = \mathbf{x} + \boldsymbol{\ell}$, and not on the absolute positions $\mathbf{x}$ and $\mathbf{x}'$ (see Figure 11.5). In particular, we have

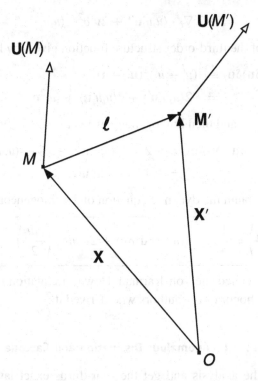

Figure 11.5 In homogeneous turbulence only the relative difference in position between the points M and M' is relevant.

$$\langle \partial_k(u_k u_i u_j') \rangle = -\partial_{\ell_k} \langle u_k u_i u_j' \rangle, \tag{11.16}$$

$$\langle \partial_k'(u_k' u_j' u_i) \rangle = \partial_{\ell_k} \langle u_k' u_j' u_i \rangle. \tag{11.17}$$

Hence we have the expression

$$\partial_t \langle u_i u_j' \rangle = -\partial_{\ell_k} \langle u_k' u_j' u_i - u_k u_i u_j' \rangle + \partial_{\ell_i} \langle P u_j' \rangle - \partial_{\ell_j} \langle P' u_i \rangle + 2\nu \, \partial^2_{\ell_k \ell_k} \langle u_i u_j' \rangle. \tag{11.18}$$

We now restrict the discussion to the trace of the second-order correlation tensor. In this case, the equation is simplified for two reasons. On the one hand, the contribution of the pressure disappears by virtue of the incompressibility,

$$\partial_{\ell_i} \langle P u_i' \rangle = \langle \partial_i'(P u_i') \rangle = \langle P \partial_i' u_i' \rangle = 0, \tag{11.19}$$

and on the other hand, we have by homogeneity

$$\partial_{\ell_k} \langle u_k u_i u_i' \rangle (\boldsymbol{\ell}) \equiv \partial_{\ell_k} \langle u_k(\mathbf{x}) u_i(\mathbf{x}) u_i(\mathbf{x} + \boldsymbol{\ell}) \rangle$$

$$= \partial_{\ell_k} \langle u_k(\mathbf{x} - \boldsymbol{\ell}) u_i(\mathbf{x} - \boldsymbol{\ell}) u_i(\mathbf{x}) \rangle \equiv \partial_{\ell_k} \langle u_k' u_i' u_i \rangle (-\boldsymbol{\ell})$$

$$= -\partial_{\ell_k} \langle u_k' u_i' u_i \rangle (\boldsymbol{\ell}). \tag{11.20}$$

Hence we have the expression

$$\partial_t \langle u_i u_i' \rangle = -2 \, \partial_{\ell_k} \langle u_k' u_i' u_i \rangle + 2\nu \, \partial^2_{\ell_k \ell_k} \langle u_i u_i' \rangle$$

$$= -2\,\nabla_\ell \cdot \langle u_i u_i' \mathbf{u}' \rangle + 2\nu\,\partial^2_{\ell_k \ell_k} \langle u_i u_i' \rangle\,. \tag{11.21}$$

The introduction of the third-order structure function gives (by homogeneity)

$$\langle (\delta \mathbf{u} \cdot \delta \mathbf{u}) \delta \mathbf{u} \rangle \equiv \langle (u_i' - u_i)^2 (\mathbf{u}' - \mathbf{u}) \rangle$$
$$= -2 \langle u_i u_i' \mathbf{u}' \rangle + 2 \langle u_i u_i' \mathbf{u} \rangle + \langle u_i^2 \mathbf{u}' \rangle - \langle u_i'^2 \mathbf{u} \rangle\,. \tag{11.22}$$

By incompressibility, and then homogeneity, we get

$$\nabla_\ell \cdot \langle (\delta \mathbf{u} \cdot \delta \mathbf{u}) \delta \mathbf{u} \rangle = -2\,\nabla_\ell \cdot \langle u_i u_i' \mathbf{u}' \rangle + 2\,\nabla_\ell \cdot \langle u_i u_i' \mathbf{u} \rangle$$
$$= -4\,\nabla_\ell \cdot \langle u_i u_i' \mathbf{u}' \rangle\,. \tag{11.23}$$

Hence finally, we obtain the dynamic equation of a homogeneous turbulence:

$$\boxed{\partial_t \left\langle \frac{u_i u_i'}{2} \right\rangle = \frac{1}{4}\,\nabla_\ell \cdot \langle (\delta \mathbf{u} \cdot \delta \mathbf{u}) \delta \mathbf{u} \rangle + 2\nu\,\partial^2_{\ell_k \ell_k} \left\langle \frac{u_i u_i'}{2} \right\rangle\,.} \tag{11.24}$$

This expression is called the von Kármán–Howarth equation (von Kármán and Howarth, 1938) in honor of the authors who derived it.[3]

11.3.2   Anomalous Dissipation and Cascade

To go further in the analysis and get the four-thirds exact law, it is necessary to introduce additional properties of turbulence. Since we make the assumption of homogeneity, we are interested only in the statistical properties of turbulence away from the boundary conditions (the walls of an experiment). For simplicity, we consider a triply periodic geometry; this geometry is also very convenient for direct numerical simulations that use the Fourier decomposition of the velocity (numerical spectral codes). In the inviscid limit, we have seen that the fluid equations conserve energy (see Eq. (3.22)) and that

$$\frac{dE}{dt} = \frac{d\langle \mathbf{u}^2/2 \rangle}{dt} = -\nu \langle \mathbf{w}^2 \rangle\,, \tag{11.25}$$

where $\langle\,\rangle$ means, for example, the average over a periodic box (the ergodic hypothesis is used because of the assumption of homogeneity). In particular, we see that the averaged energy dissipation is zero when $\nu = 0$.

Expression (11.25) brings to the fully developed turbulence theory the most important result in the limit of an infinitely large Reynolds number $R_e \to +\infty$ (which is equivalent to the limit $\nu \to 0$). The **mean rate of energy dissipation** (per unit mass) reaches a finite limit independent of the viscosity ν, traditionally denoted ε; formally, this means that

$$\boxed{\lim_{\nu \to 0} \frac{dE}{dt} = -\varepsilon \neq \left.\frac{dE}{dt}\right|_{\nu=0}\,,} \tag{11.26}$$

---

[3] In fact, the relation derived was slightly different because it involved longitudinal and transverse correlations.

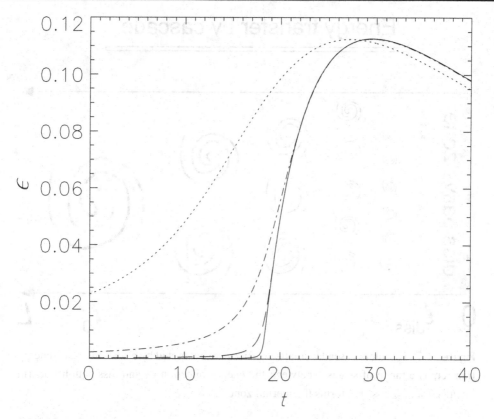

Figure 11.6 Temporal variation of the mean rate of energy dissipation ε for various Reynolds numbers: $R_e = 10^2$ (dashed line), 10^3 (dash–dot line), 10^4 (long dashes) and 10^5 (solid line). Freely decaying numerical simulation.

which is the essence of the **anomalous dissipation**. To illustrate this property, we show in Figure 11.6 the temporal variation of the mean rate of energy dissipation for a decaying turbulence (i.e. without the application of an external force) and for various Reynolds numbers. We see that $\varepsilon(t)$ tends asymptotically (at large Reynolds numbers) to a viscosity-independent form. In the presence of a stationary external force, the turbulent system adjusts so that the rates of energy injection and of energy dissipation compensate. Then, ε becomes independent of time.

Physically, the existence of a finite limit for ε when $\nu \to 0$ means that the turbulent fluid should develop structures at increasingly small scales to increase the vorticity modulus. This process of energy transfer from large to small scales can be found in the simulation where the asymptotic form of $\varepsilon(t)$ is characterized by an initial phase for which no dissipation is visible. The reason is that, in this simulation, the energy is originally put at the largest scales of the system. The initial phase therefore corresponds to the time required for that system to excite all scales – including the dissipative scales – by a process of a scale-by-scale cascade (see Figure 11.7). This image of an energy cascade was already present

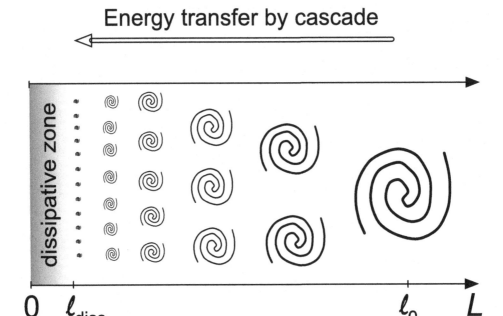

Figure 11.7 Turbulent cascade: smaller vortices are generated by a scale-by-scale energy transfer. The range of scales ℓ between the energy injection ℓ_0 and dissipation ℓ_{diss} (in practice $\ell_{\text{diss}} \ll \ell \ll \ell_0$) forms the inertial zone.

in Richardson's book published in 1922 (see Richardson (1922), p. 66) when he described qualitatively a scale-by-scale cascade of vortices, from larger to smaller scales, where finally the energy is dissipated by molecular diffusion. The intermediate range of scales on which energy is simply transferred is called the **inertial zone** and is likely to present a **universal** statistics, i.e. independent of the initial forcing and Reynolds number – provided that R_e is sufficiently large. The origin of the name inertial zone (or inertial range) comes from the fact that in this range of scales only the inertial term $(\mathbf{u} \cdot \nabla \mathbf{u})$ and pressure are important. An illustration of vortical structures present in a turbulent fluid is given in Figure 11.8: this is a direct numerical simulation[4] of the Navier–Stokes equations in which small tubes of vorticity are advected by larger tubes.

11.3.3   The Four-Thirds and Four-Fifths Exact Laws

Until now we have forgotten the external force; this is the situation described by Kolmogorov in 1941. However, in many situations an external forcing is present

---

[4] Direct numerical simulation – or simply DNS – is one of the main tools for theoreticians to investigate the turbulence properties. DNS requires very powerful computers to reach high Reynolds numbers: indeed, when R_e increases one needs to also increase the space resolution in order to correctly solve the small-scale structures. In this quest for the infinitesimal a class of models called shell models is often used (see Plunian *et al.* (2013) for a discussion in the context of MHD turbulence).

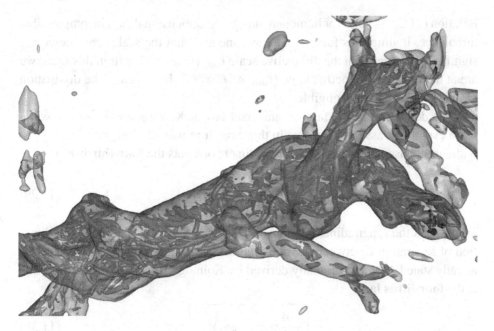

Figure 11.8 Direct numerical simulation of the Navier–Stokes equations with a space resolution of 1024^3. Two isosurfaces of vorticity are plotted in a localized region of the numerical box. (Data obtained from the JH Turbulence Database at http://turbulence.pha.jhu.edu.)

to keep the system turbulent. In this case, the stationarity assumption can be made since then the energy dissipation is statistically compensated for by the energy injection. In practice, one has to add to the Kármán–Howarth equation an extra term that we will simply denote by $\mathcal{F}$:

$$\partial_t \left\langle \frac{u_i u'_i}{2} \right\rangle = \frac{1}{4} \nabla_{\boldsymbol{\ell}} \cdot \langle (\delta u_i)^2 \, \delta \mathbf{u} \rangle + 2\nu \, \partial^2_{\ell_k \ell_k} \left\langle \frac{u_i u'_i}{2} \right\rangle + \mathcal{F}. \tag{11.27}$$

We may note that $\mathcal{F}(\boldsymbol{\ell})$ is also a correlator. We will assume that the external force acts only at the largest scales of the system and it is by nature stationary, homogeneous, and random. The assumption of stationarity allows us to eliminate the left term and say that the mean rate of energy dissipation compensates exactly – in the statistical sense – for the mean rate of energy injection. Insofar as we seek a universal law, we consider scales significantly smaller than the typical energy injection scale ℓ_0 ($\ell \ll \ell_0$); we get the relation[5]

$$0 = \frac{1}{4} \nabla_{\boldsymbol{\ell}} \cdot \langle (\delta u_i)^2 \, \delta \mathbf{u} \rangle + 2\nu \, \partial_{\ell_k \ell_k} \left\langle \frac{u_i u'_i}{2} \right\rangle + \varepsilon. \tag{11.28}$$

---

[5] A simple way to see how the force correlator contributes is to make a Taylor expansion. This is justified because the scales that interest us (the inertial zone) are assumed to be significantly smaller than those on which the force correlator varies (large-scale forcing). It yields $\mathcal{F}(\boldsymbol{\ell}) \simeq \mathcal{F}(0) + \boldsymbol{\ell} \cdot \nabla_{\boldsymbol{\ell}} \mathcal{F}(\boldsymbol{\ell})$, where $\mathcal{F}(0) = +\varepsilon$.

Relation (11.28) is valid for homogeneous, three-dimensional, and incompressible turbulence. It simplifies further if we assume also that the scales considered are significantly larger than the dissipative scale ℓ_{diss} ($\ell_{\mathrm{diss}} \ll \ell$): within this limit, we are at the heart of the inertial zone ($\ell_{\mathrm{diss}} \ll \ell \ll \ell_0$). In this case, the dissipation term in Eq. (11.28) is negligible.

The derivation of the famous universal law of Kolmogorov is now possible if we eventually assume isotropy. In this case, it remains to integrate Eq. (11.28) without the viscous term on a ball of radius ℓ; one gets the **four-thirds law**:

$$-\frac{4}{3}\varepsilon\ell = \langle(\delta u_i)^2\,\delta u_\ell\rangle, \qquad (11.29)$$

where u_ℓ is the longitudinal component of the velocity, i.e. that along the direction of separation ℓ between the two points where the measurement is made. As already stated, the law originally derived by Kolmogorov is a little different: this is the **four-fifths law**:

$$-\frac{4}{5}\varepsilon\ell = \langle(\delta u_\ell)^3\rangle, \qquad (11.30)$$

which, from the experimental point of view, is easier to measure since it requires knowledge of only one component of the velocity.[6] These results mean that the terms on the left-hand side of Eqs. (11.29) and (11.30) are negative for a direct cascade which is *a priori* a non-trivial statement. In Figure 11.9, we report schematically a series of experimental measurements of the incompressible[7] four-fifths law.

11.4   Kolmogorov Phenomenology

To derive the exact Kolmogorov law we made a rigorous analysis that is quite long. We are now going to use this prediction to build a phenomenology. A phenomenology may appear not very relevant because it deals with non-rigorous arguments. The Kolmogorov phenomenology is, however, more than a speculative proposition because it satisfies dimensionally the exact law previously derived. Nowadays, it is the Kolmogorov phenomenology which is generally used to introduce turbulence because it highlights in a simple way the main quantities that

---

[6] Contrary to appearance, the mathematical link between the four-thirds law and the four-fifths law is not trivial. Care must be taken to properly project the velocity components before calculating their statistical value.

[7] It was only in 2011 that this type of Kolmogorov law was generalized to compressible (MHD) fluids (Galtier and Banerjee, 2011; Banerjee and Galtier, 2013). Note that supersonic hydrodynamic turbulence is mostly relevant to understand the interstellar medium. In this context several important results have been obtained relatively recently thanks to three-dimensional numerical simulations (see e.g. Kritsuk *et al.* (2007) and Schmidt *et al.* (2008)).

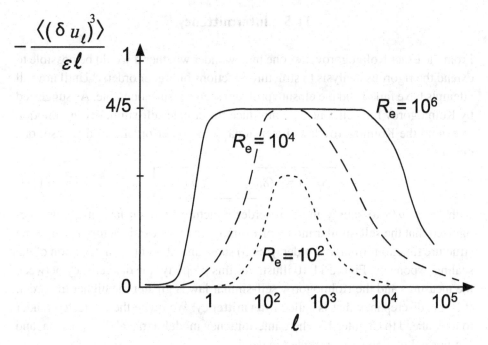

Figure 11.9 Schematic reproduction of data obtained for a decaying isotropic turbulence: the higher the Reynolds number, the better satisfied Kolmogorov's four-fifths law. Adapted from Antonia and Burattini (2006).

we need to describe the physics of turbulence. The use of this phenomenology has become so popular that sometimes one even forgets its origin. Therefore, an educated reader must not fall into this trap.

Let us define as u_ℓ the typical velocity fluctuations at scale ℓ, i.e. for example the mean quadratic difference of the velocity measured at two points separated by the distance ℓ. Our phenomenological analysis is restricted to the inertial range in which there are eddies of size ℓ. With these eddies one can associate a velocity u_ℓ and a typical eddy-turnover time $\tau_{\text{eddy}} \sim \ell/u_\ell$. In these circumstances, one can set the mean rate of energy transfer from one scale to the other as

$$\varepsilon_\ell \equiv \frac{dE_\ell}{dt} \sim \frac{u_\ell^2}{\tau_{\text{tr}}} \sim \frac{u_\ell^2}{\tau_{\text{eddy}}} \sim \frac{u_\ell^3}{\ell}, \tag{11.31}$$

where the transfer time τ_{tr} has been likened to the only time available in the inertial range, namely τ_{eddy}. Insofar as in the inertial zone the energy is neither injected nor dissipated, the mean rate of energy transfer ε_ℓ must be equal to the mean rate of energy dissipation or injection ε. This brings us to the relationship

$$\varepsilon\ell \sim u_\ell^3, \tag{11.32}$$

which is the dimensional analogue of the Kolmogorov exact law (four-fifths or four-thirds).

11.5   Intermittency

From the exact Kolmogorov law, one may wonder whether it would be possible to extend the rigorous analysis to structure functions of higher orders.[8] Until now all attempts have failed and the closure problem seems insurmountable. As suggested by Kolmogorov himself, one can introduce the simple self-similarity assumption to extend the Kolmogorov law dimensionally to higher orders. In this case, one obtains

$$S_p(\ell) \equiv \langle (\delta u_\ell)^p \rangle = C_p(\varepsilon \ell)^{\zeta_p}, \qquad (11.33)$$

with $\zeta_p = p/3$ to satisfy the third-order structure-function law. It is now recognized that the self-similar model of Kolmogorov is not satisfactory because the structure functions of higher order ($p > 3$) show unambiguously a deviation of the scaling exponents. Figure 11.10 illustrates this property: the discrepancy between the measures and the Kolmogorov self-similar law is greater the higher the order. It is this discrepancy that is called **intermittency**. We invite the interested reader to look ahead to Chapter 13 where intermittency models (fractal, log-Poisson, and log-normal models) are presented in detail.

A simple way to visualize intermittency is to plot the spatial variation of the velocity as well as its derivative. In Figure 11.11, we see schematically that the

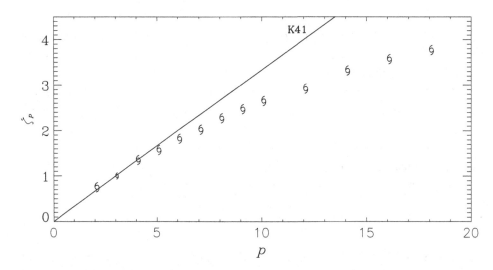

Figure 11.10 Reproduction of experimental measurements (symbols §) of the velocity structure-function exponents ζ_p (see e.g. Frisch (1995)). The self-similar Kolmogorov law (K41) is plotted for comparison. The discrepancy with the self-similar law is called intermittency.

---

[8] Remember that in general a random variable follows a probability law whose form is even better known that we can calculate its high-order moments. In our case, it is the longitudinal velocity increment δu_ℓ that serves as basic random variable.

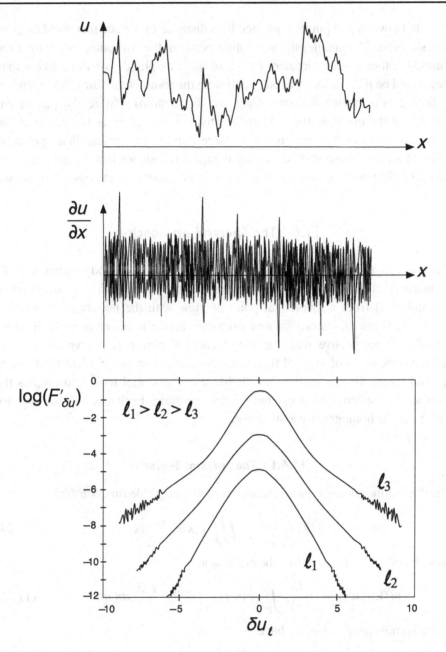

Figure 11.11 Schematic representation of the spatial variation of a turbulent velocity $u(x)$ (top), its derivative (middle), and its probability density function $F'_{\delta u}$ (bottom) for different separations ℓ. The greatest non-Gaussian wings correspond to the smallest separations.

intermittent character is amplified on the derivative with the presence of intermittent bursts, i.e. the sudden appearance of large amplitude fluctuations. A Gaussian random function has, however, the same behavior as its derivative. If now we plot the probability density function of the velocity increment (i.e. the difference in velocity between two points separated by a distance ℓ), we see that non-Gaussian wings appear. The smaller the separation between the two points, the larger the wings. In other words, turbulent events of large amplitude are more likely than they would be if the velocity at one point were the result of the sum of independent random events (which follow a Gaussian distribution). Physically, this means that in a turbulent flow the velocity fluctuation at a point is the result of the superimposition of the influence of a large number of vortices that operate at different scales. These vortices are not completely independent of each other but have a spatio-temporal memory whose origin lies mainly in the cascade process.

11.6 The Spectral Approach

The spectral approach provides a complementary way to study turbulence. To be more precise, we can say that the use of the spectral (i.e. Fourier) space is justified (i) from a theoretical point of view with the possibility to analyze the mechanisms of interaction and exchange between wavenumbers, (ii) from a modeling perspective with the introduction of new closure hypotheses, and (iii) from the point of view of the numerical simulation of turbulent fields using spectral codes. In this section, we define a few essential tools and discuss the most studied spectral quantity: the energy spectrum. In that which follows, we will make the homogeneity assumption.

11.6.1 The Spectral Tensor

The three-dimensional Fourier transform of an integrable function $f(\mathbf{x})$ is

$$\hat{f}(\mathbf{k}) \equiv \frac{1}{(2\pi)^3} \iiint f(\mathbf{x})e^{-i\mathbf{k}\cdot\mathbf{x}} \, d\mathbf{x} \,. \tag{11.34}$$

From this definition, we define the correlator:

$$\langle \hat{f}(\mathbf{k})\hat{f}(\mathbf{k}')\rangle = \frac{1}{(2\pi)^6} \int_{\mathbf{R}^6} \langle f(\mathbf{x})f(\mathbf{x}')\rangle e^{-i(\mathbf{k}\cdot\mathbf{x}+\mathbf{k}'\cdot\mathbf{x}')} \, d\mathbf{x} \, d\mathbf{x}' \,. \tag{11.35}$$

In the homogeneous case, we have

$$\langle f(\mathbf{x})f(\mathbf{x}')\rangle = \langle f(\mathbf{x})f(\mathbf{x}+\boldsymbol{\ell})\rangle = Q(\boldsymbol{\ell}) \,, \tag{11.36}$$

which leads to

$$\begin{aligned}
\langle \hat{f}(\mathbf{k})\hat{f}(\mathbf{k}')\rangle &= \frac{1}{(2\pi)^6} \int_{\mathbf{R}^6} Q(\boldsymbol{\ell})e^{-i(\mathbf{k}+\mathbf{k}')\cdot\mathbf{x}}e^{-i\mathbf{k}'\cdot\boldsymbol{\ell}} \, d\mathbf{x} \, d\boldsymbol{\ell} \\
&= \frac{1}{(2\pi)^3} \iiint Q(\boldsymbol{\ell})e^{-i\mathbf{k}'\cdot\boldsymbol{\ell}}\delta(\mathbf{k}+\mathbf{k}')d\boldsymbol{\ell} \\
&= \delta(\mathbf{k}+\mathbf{k}')\hat{Q}(\mathbf{k}') \,.
\end{aligned} \tag{11.37}$$

In hydrodynamic turbulence, we are mainly interested in the velocity field.[9] Therefore, we construct the three-dimensional spectral tensor of the velocity field for a homogeneous turbulence:

$$\Phi_{ij}(\mathbf{k}) \equiv \frac{1}{(2\pi)^3} \iiint R_{ij}(\boldsymbol{\ell}) e^{-i\mathbf{k}\cdot\boldsymbol{\ell}} \, d\boldsymbol{\ell} \,, \tag{11.38}$$

where $R_{ij}(\boldsymbol{\ell}) \equiv \langle u_i(\mathbf{x})u_j(\mathbf{x}+\boldsymbol{\ell})\rangle$ is the second-order velocity correlation tensor. This quantity is well defined mathematically (it is an integrable function) to the extent that the correlation between two points tends to zero when the distance between these points increases to infinity. We have the hermitian property

$$\Phi_{ji}(\mathbf{k}) = \Phi_{ij}(-\mathbf{k}) = \Phi_{ij}^*(\mathbf{k}) \,, \tag{11.39}$$

where * means the complex conjugate. This property is in particular the result of the homogeneity relationship in the physical space: $R_{ij}(\boldsymbol{\ell}) = R_{ji}(-\boldsymbol{\ell})$. Additionally, the incompressible assumption gives

$$k_i \Phi_{ij}(\mathbf{k}) = 0 \,. \tag{11.40}$$

11.6.2 The Energy Spectrum

We will now define one of the best-studied turbulence quantities, namely the kinetic energy spectrum per unit mass. We have

$$E = \frac{1}{2}\langle u^2 \rangle = \frac{1}{2} R_{ii}(0) = \frac{1}{2} \iiint \Phi_{ii}(\mathbf{k}) d\mathbf{k} \,. \tag{11.41}$$

This relation allows us to define the **energy spectrum**:

$$E(\mathbf{k}) \equiv \frac{1}{2}\Phi_{ii}(\mathbf{k}) \,, \tag{11.42}$$

which reflects the distribution of energy by spectral band $[k, k+dk]$. In the particular case of isotropic turbulence, the spectral tensor depends only on the modulus of $\mathbf{k}$, also the definition of the spectrum is reduced to

$$E(k) = \iint E(\mathbf{k}) dS(\mathbf{k}) = 2\pi k^2 \Phi_{ii}(k) \,, \tag{11.43}$$

where $S(\mathbf{k})$ is a sphere of radius $\mathbf{k}$ in the Fourier space. The energy of the system is found by a simple integration over k:

$$E = \int_0^{+\infty} E(k) dk \,. \tag{11.44}$$

---

[9] Generally, in turbulence the velocity field does not decrease rapidly to infinity and the Fourier transform can be applied only in the framework of the theory of distributions (i.e. generalized functions). The situation is different for the second-order velocity correlation tensor (see Section 11.2).

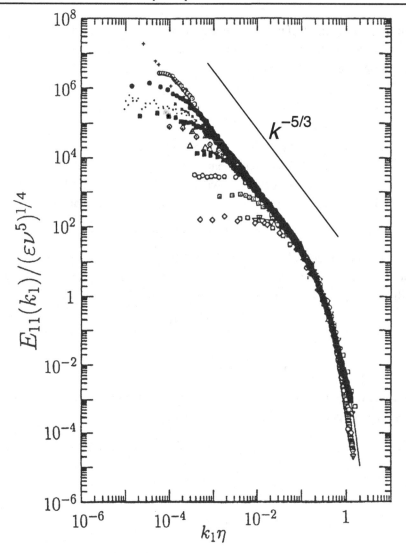

Figure 11.12 The energy spectrum found in experimental measurements: a universal $k^{-5/3}$ law is identified over approximately four orders of magnitude. From Saddoughi and Veeravalli (1994).

11.6.3   The Kolmogorov $k^{-5/3}$ Spectrum

The Kolmogorov phenomenology – which is valid for isotropic turbulence – gave us the dimensional relationship $\varepsilon\ell \sim u_\ell^3$. From this expression, we can deduce dimensionally

$$u_\ell^2 \sim (\varepsilon\ell)^{2/3} \sim \varepsilon^{2/3} k^{-2/3} \sim E(k)k, \tag{11.45}$$

hence the spectral prediction (one introduces the equals sign and therefore a constant of proportionality C_K)

$$E(k) = C_K \varepsilon^{2/3} k^{-5/3} , \tag{11.46}$$

where C_K is the Kolmogorov constant whose latest measurements give $C_K \simeq 2$. Although it was deduced for the first time (independently) by Obukhov and Heisenberg (Heisenberg, 1948), this spectrum is known as the **Kolmogorov spectrum** (denoted, in an abuse of notation, K41). It is quite remarkable that many laboratory experiments have very well confirmed this phenomenological law over several orders of magnitude as shown in Figure 11.12.

Using our analysis in the physical space and the Kolmogorov phenomenology, we may interpret Figure 11.12 in the following manner. The $k^{-5/3}$ spectrum is the result of a balance between energy injection at small wavenumber and its dissipation at large wavenumber. Between the two, we have an inertial range where turbulence behaves in a universal way. This inertial range is crossed by an energy flux that connects the large scales to the small scales: this is the region where the energy cascades. The cascade process means that the energy flows scale-by-scale in a continuous manner. The Kolmogorov spectrum thus consists not of a set of lines but of a continuous distribution of energy. Contrary to what is assumed in the Kolmogorov phenomenology, we now know that the energy flux is not exactly constant in the inertial range. These flux fluctuations are actually at the origin of anomalous exponents of the structure functions (i.e. intermittency).

One of the objectives of theoreticians is to find a rigorous justification for the Kolmogorov spectrum. This requires one to obtain a self-consistent dynamic equation for the energy spectrum. Despite numerous attempts, so far no solution to this problem has been found by physicists and mathematicians. The fundamental reason is related to a closure problem whose origin is the non-linearity of the Navier–Stokes equations. We will return in Chapter 13 to this closure problem in the spectral space which finds a particular exact solution in incompressible MHD: this is the anisotropic regime of weak Alfvén wave turbulence.

MHD Turbulence

The objective of this chapter is to present the main properties of MHD turbulence. This presentation will be based on the concepts introduced in the previous chapter on hydrodynamics. Indeed, historically the progress made on conducting fluids was often the consequence of the direct application or adaptation of ideas developed originally in the hydrodynamic community. We will see, however, that the MHD community has its own identity, since specific concepts for conducting fluids have been successfully introduced.

12.1 From Astrophysics to Tokamaks

12.1.1 Solar Wind

We begin this chapter by presenting three examples of turbulent plasmas borrowed from astrophysics where turbulence is ubiquitous. The first example is that of the solar wind – the only astrophysical environment where *in situ* measurements have been possible since the space age in the 1960s. In Figure 12.1 we see the spectrum of the interplanetary magnetic field fluctuations carried by the solar wind. It is a spectrum of frequency f because the measures are based on a temporal signal. Insofar as the solar wind propagates relatively quickly (its speed U_{SW} varies between 300 km/s and 1000 km/s), we can assume that the recorded temporal signal gives a fairly accurate picture of the plasma at a given moment: the plasma remains relatively fixed for the duration of the measurement and therefore we have $k \simeq 2\pi f / U_{SW}$ (this is Taylor's hypothesis). In other words, the frequency spectrum can be interpreted in a first approximation as a wavenumber spectrum. This spectrum shows three power laws. At very low frequencies ($f < 10^{-4}$ Hz), we have a f^{-1} law whose origin is generally attributed to the physical processes of the low solar corona: this frequency range can therefore be interpreted as the area of energy injection for the heliospheric turbulence. For higher frequencies,

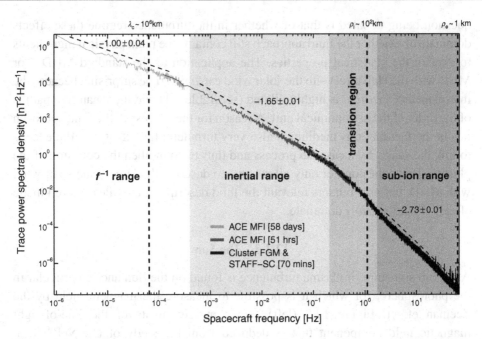

Figure 12.1 Frequency spectrum of the magnetic fluctuations measured in the ecliptic solar wind at 1 AU with Cluster/ESA and ACE/NASA. Three power laws can be distinguished according to the scales. The vertical dashed lines indicate the correlation length scale λ_c below which the plasma fluctuations have lost memory of their solar origins, the ion gyro-radius ρ_i, and the electron gyro-radius ρ_e. Courtesy of Dr. K. H. Kiyani; see also Kiyani *et al.* (2015).

i.e. up to 0.5 Hz, a second power law is observed in $f^{-5/3}$: this range of frequencies is interpreted as an MHD inertial range where additionally Alfvén waves are detected. Finally, a third power law emerges at frequencies between 0.5 Hz and 50 Hz: this is essentially a dispersive inertial range where a significantly steeper spectrum varying as $f^{-8/3}$ is measured (Kiyani *et al.*, 2009; Alexandrova *et al.*, 2012). It is apparent that an additional physical ingredient to the standard MHD model is necessary in order to reproduce such a law. This ingredient is linked to the decoupling between the ions and electrons: with Taylor's hypothesis we see that the spatial scale associated with a frequency of 1 Hz is of the order of the ion inertial length (which is also of the order of the ion gyro-radius). Therefore, the fluid approach adapted to the second inertial zone is the Hall MHD. It can be shown theoretically and numerically that actually a spectrum between $-7/3$ and -3 is expected when the Hall dispersive effects are taken into account, which corresponds approximately to the range of values observed. Beyond 50 Hz, the spectrum stiffens again: one reaches the electronic scales at which the current instruments saturate. At these scales, and even once $f > 1$ Hz, kinetic effects are involved in the dynamics of the heliospheric plasma. The currently fundamental question being debated is that of whether in the turbulence regime these effects dominate or whether the fluid approach still contains the main physical ingredients to explain the statistical properties. The application of the standard MHD – or MHD with the Hall effect – to the solar wind can *a priori* be surprising because the interplanetary medium is highly diluted (see Table 1.1) and the mean free path is of the order of the astronomical unit. A reason for the success of the fluid approach lies in the fact that this medium is also very turbulent: the effect of turbulence is to link the scales by a cascade process and thus to strengthen the cohesion of the plasma. This cohesion not only allows one to describe the large-scale solar wind with MHD, but also renders relevant the fluid description at scales where kinetic effects should *a priori* dominate.

12.1.2   The Sun

A second signature of plasma turbulence is found on the Sun and in particular in the photosphere, for which it is possible to measure the magnetic field by the Zeeman effect. Figure 12.2 (left) shows measurements of the line-of-sight magnetic field component that is deduced from the study of the Ni I line at 676.8 nm. In the plot on the right, intermittency is found in active regions. The study demonstrates that intermittency tends to be more pronounced when the region is more active – that is, the flux emitted by the region in the X-ray domain

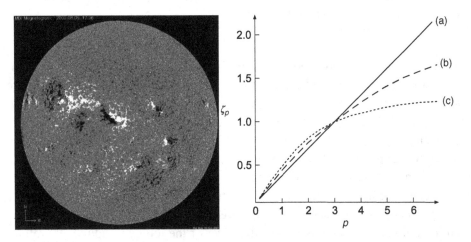

Figure 12.2 Left: The line-of-sight photospheric magnetic field (inward field in black, outward field in white), August 9, 2002. Active regions are characterized by a concentration of strong magnetic fields (like e.g. in the middle of the image). Credit: MDI/SOHO/NASA. Right: Schematic view of the structure-function exponents ζ_p of the corresponding magnetic field components for different active regions (from the less active (a) to the more active (c)). Intermittency increases with the activity of the region which is measured through the X-ray flux. See e.g. Abramenko *et al.* (2002).

is important – and it disappears after the triggering of a solar flare. Thus a solar flare leads to relaxation of the system. This type of study opens the way to an emerging field: that of space weather, i.e. the science which aims to predict major solar events that could impact the Earth with all the consequences that we can imagine (disturbance or destruction of satellites, a critical increase in radiation at high altitude with an impact on flights, over voltages or even power cuts on Earth).

12.1.3 The Interstellar Medium

A third well-identified signature of plasma turbulence is given by the interstellar medium (Figure 12.3). Historically, the first observational signatures of interstellar turbulence date back to the 1950s (Münch, 1958) when spectral line broadening via Doppler shifts was detected. In particular, the spectral line widths found in cold molecular clouds are much broader than would be expected from the low gas temperatures ($T \sim 10$ K), and the velocities far exceed the speed of sound ($c_S \sim 0.5$ km/s). More recent studies on rotational lines of some molecules[1] in the metric domain have significantly extended the first measurements. Figure 12.4 gives a schematic illustration with nearly 30 molecular clouds (in our galaxy): the velocity difference (velocity increment) between two points is shown as a

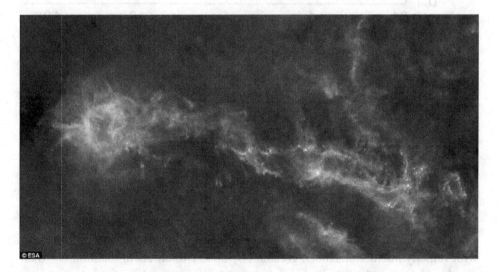

Figure 12.3 Interstellar filaments, at a distance from Earth of 500 pc, observed in the infrared domain (at 70, 250, and 500 μm) using the European Herschel space telescope. The stars are born along these filaments. Credits: ESA/Herschel/SPIRE/PACS.

---

[1] Molecular clouds are efficient radiators at infrared and radio wavelengths, which allows us to get physical information about e.g. velocity fluctuations, temperatures, or magnetic fields. Information is obtained mainly from line radiation emitted by minority molecules such as CO, OH, and CN or more complex molecules (Biskamp, 2003).

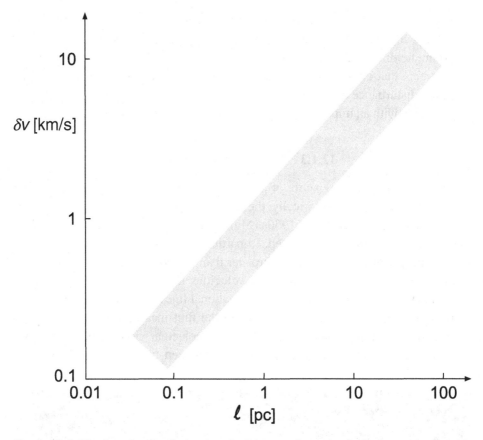

Figure 12.4 The domain of variation (gray) of the non-thermal velocity increment δv with the scale ℓ. Measurements made in 27 interstellar clouds (in our galaxy) from the emission of the molecular line $^{12}CO\ J = 1-0$. A power law emerges, with $\zeta_1 = 0.56 \pm 0.02$. Adapted from Heyer and Brunt (2004).

function of ℓ, the distance between these points. A $\langle \delta v(\ell) \rangle \sim \ell^{0.56}$ law emerges over three orders of magnitude (between 0.04 pc and 40 pc; 1 pc $\simeq 3 \times 10^{16}$ m). Dimensionally, this scaling law is compatible with a velocity fluctuation spectrum of $E(k) \sim k^{-2.12}$. Moreover, these measurements show a supersonic turbulence with velocity fluctuations much greater – by up to two orders of magnitude – than the thermal velocities: then, the turbulent Mach number based on the velocity fluctuations is very often greater than 10.

Imaging allows one also to characterize the interstellar medium and show that this environment comes in the form of huge networks of filaments in which stars are born (Figure 12.3). The most recent measurements with the space Herschel/ESA telescope (Arzoumanian *et al.*, 2011) seem to reveal that these filaments, which span tens of parsecs, have the same width ($\sim$1 pc) regardless of the density or the length of these filaments. This characteristic scale might correspond to the sonic scale, i.e. the scale at which interstellar turbulence goes from a supersonic regime to a subsonic one. Compressible MHD turbulence in fact appears to be one of the key ingredients to understand the dynamics of interstellar clouds and its consequences for, for example, the rate of star formation (Chabrier and Hennebelle, 2011).

Whether to understand the dynamics of the Sun and the solar wind, or star formation in the interstellar medium, it seems essential to take turbulence into account. This requires a minimum understanding of turbulent processes operating in plasmas. In this chapter, we present the main properties of incompressible MHD turbulence.

12.1.4 Tokamaks

Laboratory plasmas like in tokamaks are the subject of severe turbulence affecting their confinement. To fight against the development of such turbulence, it is important to learn more about plasma and hence to measure as precisely as possible its characteristics. The diagnostic methods used for these fusion plasmas are quite different from those used for space plasmas because it is preferable to opt for non-intrusive methods. Thus, the technique of scattering of electromagnetic waves by density fluctuations is often used to infer, for example, the spectrum of density fluctuations. Figure 12.5 shows (schematically) an example of measurements in Tore Supra: this is the turbulent spectrum of density fluctuations. The plateau at small wavenumber is interpreted as the zone of energy injection where primary instabilities develop. At larger wavenumber, we have an inertial zone where energy is transferred from large to small scales. The nature of turbulence in fusion plasmas is complex because on the one hand it develops mainly near the walls of the reactor, and therefore the assumption of homogeneity is not well verified, and on the other hand it involves processes that go well beyond the MHD approximation (Diamond *et al.*, 2010). However, the concepts used to attack the turbulence problem in tokamaks are fundamentally based on those we are discussing in this part of the book.

12.2 Exact Laws

12.2.1 Four-Thirds Law

In this section, we shall establish the four-thirds law by using the calculation steps introduced in the hydrodynamic case. The incompressible standard MHD equations are

$$\partial_t u_i + \partial_k (u_k u_i) = -\partial_i P_* + \partial_k (b_k b_i) + \nu\, \partial_{kk} u_i + f_i, \tag{12.1}$$

$$\partial_t b_i + \partial_k (u_k b_i) = \partial_k (b_k u_i) + \eta\, \partial_{kk} b_i. \tag{12.2}$$

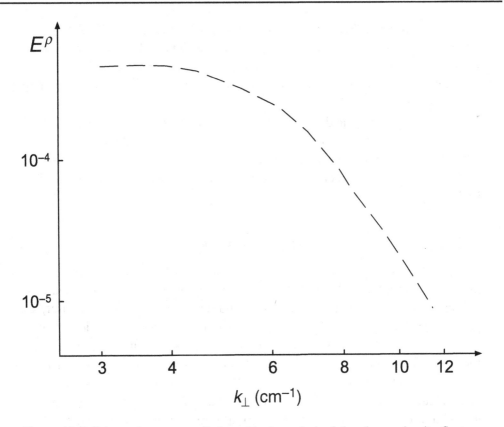

Figure 12.5 Schematic spectrum (in logarithmic scales) of the plasma density fluctuations produced by the Tore Supra tokamak in Cadarache. Adapted from Vermare *et al.* (2011).

For homogeneous MHD turbulence, we obtain the following dynamic equations for the kinetic and magnetic second-order correlation tensors (we forget for the moment the external force; see Section 11.3.1 for a detailed calculation):

$$\partial_t \langle u_i u_j' \rangle = -\partial_{\ell_k} \langle u_k' u_j' u_i - u_k u_i u_j' \rangle + \partial_{\ell_k} \langle b_k' b_j' u_i - b_k b_i u_j' \rangle$$
$$+ \partial_{\ell_i} \langle P_* u_j' \rangle - \partial_{\ell_j} \langle P_*' u_i \rangle + 2\nu \, \partial^2_{\ell_k \ell_k} \langle u_i u_j' \rangle \,, \tag{12.3}$$

$$\partial_t \langle b_i b_j' \rangle = -\partial_{\ell_k} \langle u_k' b_j' b_i - u_k b_i b_j' \rangle + \partial_{\ell_k} \langle b_k' u_j' b_i - b_k u_i b_j' \rangle$$
$$+ 2\eta \partial^2_{\ell_k \ell_k} \langle b_i b_j' \rangle \,. \tag{12.4}$$

By summing the trace of tensors, we obtain after simplification

$$\partial_t \langle u_i u_i' + b_i b_i' \rangle = -2 \, \partial_{\ell_k} \langle u_k' u_i' u_i + u_k' b_i' b_i \rangle + 2 \, \partial_{\ell_k} \langle b_k' b_i' u_i + b_k' u_i' b_i \rangle$$
$$+ 2\nu \, \partial^2_{\ell_k \ell_k} \langle u_i u_i' \rangle + 2\eta \, \partial^2_{\ell_k \ell_k} \langle b_i b_i' \rangle \,. \tag{12.5}$$

The introduction of structure functions gives in particular

$$\nabla_\ell \cdot \langle (\delta \mathbf{u} \cdot \delta \mathbf{u}) \delta \mathbf{u} \rangle = -4 \, \nabla_\ell \cdot \langle u_i u_i' \mathbf{u}' \rangle \,, \tag{12.6}$$

$$\nabla_\ell \cdot \langle (\delta \mathbf{b} \cdot \delta \mathbf{b}) \delta \mathbf{u} \rangle = -4 \nabla_\ell \cdot \langle b_i b_i' \mathbf{u}' \rangle, \tag{12.7}$$

$$\nabla_\ell \cdot \langle (\delta \mathbf{u} \cdot \delta \mathbf{b}) \delta \mathbf{b} \rangle = -2 \nabla_\ell \cdot \langle u_i b_i' \mathbf{b}' + u_i' b_i \mathbf{b}' \rangle. \tag{12.8}$$

Hence, the generalized (MHD) von Kármán–Howarth equation is

$$\partial_t \left\langle \frac{u_i u_i' + b_i b_i'}{2} \right\rangle = \frac{1}{4} \nabla_\ell \cdot \langle (\delta \mathbf{u} \cdot \delta \mathbf{u} + \delta \mathbf{b} \cdot \delta \mathbf{b}) \delta \mathbf{u} \rangle - \frac{1}{2} \nabla_\ell \cdot \langle (\delta \mathbf{u} \cdot \delta \mathbf{b}) \delta \mathbf{b} \rangle$$

$$+ 2 \partial_{\ell_k \ell_k} \left\langle \frac{\nu u_i u_i' + \eta b_i b_i'}{2} \right\rangle. \tag{12.9}$$

The introduction of an external force and of the mean rate of total (kinetic plus magnetic) energy injection ε^T gives the following vector relationship which is valid at the heart of the inertial range (i.e. when $\ell_{\text{diss}} \ll \ell \ll \ell_0$) and for stationary turbulence:

$$0 = \frac{1}{4} \nabla_\ell \cdot \langle (\delta \mathbf{u} \cdot \delta \mathbf{u} + \delta \mathbf{b} \cdot \delta \mathbf{b}) \delta \mathbf{u} \rangle - \frac{1}{2} \nabla_\ell \cdot \langle (\delta \mathbf{u} \cdot \delta \mathbf{b}) \delta \mathbf{b} \rangle + \varepsilon^T. \tag{12.10}$$

We can integrate this equation over a ball if isotropy is assumed; we end up with the exact **four-thirds law** for MHD obtained in 1998 (Politano and Pouquet, 1998):

$$-\frac{4}{3} \varepsilon^T \ell = \langle (\delta \mathbf{u} \cdot \delta \mathbf{u} + \delta \mathbf{b} \cdot \delta \mathbf{b}) \delta u_\ell \rangle - 2 \langle (\delta \mathbf{u} \cdot \delta \mathbf{b}) \delta b_\ell \rangle, \tag{12.11}$$

where u_ℓ and b_ℓ are the longitudinal components of the velocity and magnetic field, i.e. those along the direction of separation ℓ between the two points.

12.2.2 Elsässer Variables and Exact Non-linear Solution

Until now, we have studied the MHD equations through the velocity and magnetic field (which can be normalized with respect to a velocity in the incompressible case). We can introduce other variables combining the two fields: these are the **Elsässer variables**, which are particularly interesting in the incompressible case. They were proposed by W. Elsässer (1950) and are defined as

$$\mathbf{z}^\pm \equiv \mathbf{u} \pm \mathbf{b}. \tag{12.12}$$

If now one rewrites the MHD equations, one obtains

$$\nabla \cdot \mathbf{z}^\pm = 0, \tag{12.13}$$

$$\frac{\partial \mathbf{z}^+}{\partial t} + \mathbf{z}^- \cdot \nabla \mathbf{z}^+ = -\nabla P_* + \nu_+ \Delta \mathbf{z}^+ + \nu_- \Delta \mathbf{z}^-, \tag{12.14}$$

$$\frac{\partial \mathbf{z}^-}{\partial t} + \mathbf{z}^+ \cdot \nabla \mathbf{z}^- = -\nabla P_* + \nu_- \Delta \mathbf{z}^+ + \nu_+ \Delta \mathbf{z}^-, \tag{12.15}$$

where $v_\pm \equiv (v \pm \eta)/2$. Under this form, the MHD equations appear more symmetrical. Even more interesting is to note that $\mathbf{z}^+ = \mathbf{0}$ (or $\mathbf{z}^- = \mathbf{0}$) is an exact non-linear solution of the equations. Indeed, then we have

$$\mathbf{z}^\pm = \mathbf{0} \quad \Rightarrow \quad \frac{\partial \mathbf{z}^\mp}{\partial t} = \eta \, \Delta \mathbf{z}^\mp . \tag{12.16}$$

Therefore, the evolution of the solution is done on a dissipative time scale, which can be very long if the associated Reynolds (kinetic and magnetic) numbers are very large – as in astrophysics. Finally, note that the introduction of the Elsässer variables raises a serious mathematical problem: these variables are the addition of a true vector, the velocity, and a pseudo-vector, the magnetic field. What we do here would be considered audacious by mathematicians. In particular, the use of the tensorial analysis must be done with caution because the symmetries satisfied by true and pseudo-vectors are not the same.

12.2.3 Return to the Four-Thirds Law

The exact four-thirds law can also be written with the Elsässer variables. Following a similar approach, it can easily be shown that

$$-\frac{4}{3}\varepsilon^\pm \ell = \langle (\delta \mathbf{z}^\pm \cdot \delta \mathbf{z}^\pm) \delta z_\ell^\mp \rangle , \tag{12.17}$$

where $\varepsilon^\pm$ is the mean rate of Elsässer energy injection. We obtain two universal laws instead of one as previously. The reason is that the combination of the velocity and magnetic field allows us to build another inviscid and ideal invariant, the cross-helicity, for which we can also get a universal law.[2] With these variables, the two laws are calculated at the same time. To conclude, note that the exact relation (12.17) was used to demonstrate the presence of an inertial energy cascade in the solar wind (Sorriso-Valvo *et al.*, 2007). Indeed, from the solar wind data it is possible to show the existence of a linear relation for the third-order structure function. By this method one can obtain an estimate of $\varepsilon^\pm$ which can also be interpreted as the mean rate of energy dissipation.

12.3 Iroshnikov–Kraichnan Phenomenology

12.3.1 Alfvén Wave-Packets

In the presence of a uniform magnetic field $\mathbf{b}_0$, we obtain

$$\frac{\partial \mathbf{z}^\pm}{\partial t} \mp \mathbf{b}_0 \cdot \nabla \mathbf{z}^\pm + \mathbf{z}^\mp \cdot \nabla \mathbf{z}^\pm = -\nabla P_* + v_+ \, \Delta \mathbf{z}^\pm + v_- \, \Delta \mathbf{z}^\mp . \tag{12.18}$$

---

[2] The exact law for the cross-helicity is $-\frac{4}{3}\varepsilon^C \ell = -\langle (\delta \mathbf{u} \cdot \delta \mathbf{u} + \delta \mathbf{b} \cdot \delta \mathbf{b}) \delta b_\ell \rangle + 2\langle (\delta \mathbf{u} \cdot \delta \mathbf{b}) \delta u_\ell \rangle$, where ε^C is the mean rate of cross-helicity injection (Politano and Pouquet, 1998).

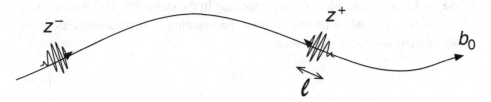

Figure 12.6 Propagation of two Alfvén wave-packets along a quasi-uniform magnetic field.

The variables $\mathbf{z}^{\pm}$ can then be assimilated with **Alfvén wave-packets** propagating along the uniform magnetic field at the Alfvén speed $\mathbf{b}_0$ (Figure 12.6). An Alfvén wave-packet undergoes a non-linear deformation when it collides with another wave-packet propagating in the opposite direction. This property is important e.g. for the solar wind in which the flux of Alfvén waves escaping the Sun is significantly larger than the inward flux. In the context of incompressible MHD, one sees that turbulence can be fully developed only if there is enough inward/outward wave flux.

12.3.2   The Energy Spectrum in $k^{-3/2}$

The discussion on the Alfvén wave-packets leads us to introduce one of the major ideas of MHD turbulence: the IK phenomenology proposed independently by Iroshnikov (1964) and Kraichnan (1965).

The main difference between incompressible hydrodynamics and MHD is the presence of waves in the latter case. We can associate with these waves an Alfvén time, τ_A, that characterizes the duration of interaction – the collision – between two wave-packets. If these two wave-packets have a typical length ℓ (see Figure 12.6), then

$$\tau_A \sim \frac{\ell}{b_0}, \tag{12.19}$$

where b_0 is the Alfvén speed. The main idea of the IK phenomenology is that at each scale of the inertial range the structures mainly experience a locally uniform magnetic field, even if there is no large-scale uniform magnetic field. In other words, even if the medium is isotropic, at the turbulent scales within the inertial range we consider that the physics is governed by the process described in Figure 12.6. Through this approach, the collision between Alfvén wave-packets becomes the central element in the analysis of MHD turbulence: it is the multiplicity of collisions that distorts the wave-packets and that eventually transfers energy towards smaller scales.

We will follow the Kolmogorov phenomenology and assume that turbulence is isotropic. Because we have introduced the concept of a wave-packet, we have to use the Elsässer variables. We are operating in the symmetrical situation where $z^+ \sim z^- \sim z$, i.e. when the wave fluxes propagating in opposite directions are approximately the same. Therefore

$$\varepsilon \sim \frac{z_\ell^2}{\tau_{tr}}, \qquad (12.20)$$

is the energy transfer time from one scale to the other in the inertial range. In the process described, turbulence is supposed to grow following many stochastic collisions between Alfvén wave-packets, and at a scale ℓ of the inertial range the quasi-uniform magnetic field is assumed to be significantly larger than the magnetic fluctuations which means that $\tau_{tr} \gg \tau_A$. To estimate the transfer time, let us first evaluate the wave-packet deformation at the scale ℓ produced by a single collision (we are on the wave-packet); we get

$$z_\ell(t + \tau_A) \sim z_\ell(t) + \tau_A \frac{\partial z_\ell}{\partial t} \sim z_\ell(t) + \tau_A \frac{z_\ell^2}{\ell}. \qquad (12.21)$$

Therefore, the deformation of the wave-packet after one collision is

$$\Delta_1 z_\ell \sim \tau_A \frac{z_\ell^2}{\ell}. \qquad (12.22)$$

This deformation will grow with time and for N stochastic collisions the cumulative effect can be evaluated in the same manner as a random walk:

$$\sum_{i=1}^{N} \Delta_i z_\ell \sim \tau_A \frac{z_\ell^2}{\ell} \sqrt{\frac{t}{\tau_A}}. \qquad (12.23)$$

The transfer time we are looking for is that for which the cumulative deformation is of order one, i.e. of the order of the wave-packet itself:

$$z_\ell \sim \sum_{1}^{N} \Delta_i z_\ell \sim \tau_A \frac{z_\ell^2}{\ell} \sqrt{\frac{\tau_{tr}}{\tau_A}}. \qquad (12.24)$$

One then obtains

$$\tau_{tr} \sim \frac{1}{\tau_A} \frac{\ell^2}{z_\ell^2} \sim \frac{\tau_{eddy}^2}{\tau_A}, \qquad (12.25)$$

where τ_{eddy} is the eddy-turnover time that we introduced in hydrodynamic turbulence. It remains to complete the calculation using expression (12.20); assuming $E(k) \sim E^{\pm}(k)$, one gets

$$\varepsilon \sim \frac{z_\ell^2}{\tau_{eddy}^2/\tau_A} \sim \frac{z_\ell^4}{\ell b_0} \sim \frac{E^2(k)k^3}{b_0}. \qquad (12.26)$$

Hence, the Iroshnikov–Kraichnan spectrum:

$$E(k) = C_{\text{IK}}\sqrt{\varepsilon b_0}\, k^{-3/2}, \qquad\qquad (12.27)$$

where C_{IK} is a constant of order one. This isotropic prediction differs therefore from the Kolmogorov one with a slightly less steep spectrum.[3] It is interesting to note that in this approach the transfer time is larger than it is in hydrodynamics (as $\tau_A \ll \tau_{\text{eddy}}$). Physically, it is the sporadic collisions between wave-packets – and not the continuous interactions between vortices – which explain the slow-down in the energy transfer towards the small scales. This slowdown is observed, for example, in numerical simulations where the free decay of energy is clearly slower than in hydrodynamics (Galtier *et al.*, 1997).

The IK spectrum is currently still being discussed. One reason is that the Kolmogorov spectrum in $-5/3$ is dimensionally compatible with the exact four-thirds law (12.11). Another reason is that the solar wind observations show a magnetic spectrum[4] in $-5/3$ and a velocity spectrum in $-3/2$ (Podesta *et al.*, 2007). With regard to the direct numerical simulations, the question is not yet decided because the small difference between the two predictions requires the formation of an extended (at least two orders of magnitude) inertial range in order to draw valid conlusions, which demands very large numerical resources. However, it seems that we tend to see $-5/3$ in three-dimensional isotropic simulations and $-3/2$ in three-dimensional anisotropic (i.e. with a moderate uniform magnetic field) simulations. In this case, should we take into consideration the Alfvén waves in the MHD phenomenology? If so, how? We will see below that the IK approach is essential to understanding MHD turbulence in the presence of a *real* external magnetic field. This field leads to anisotropic turbulence where actually the Alfvén wave-packets have a central role. In the limit of a strong external magnetic field, we fall within the regime of weak Alfvén wave turbulence (see Section 12.7), for which rigorous perturbative theoretical developments are possible.

12.4 Intermittency

As in hydrodynamics, intermittency is a property that has clearly been identified in MHD turbulence. Figure 12.7 (bottom) shows velocity and magnetic field fluctuation measurements made in the solar wind at low frequencies. We see that

---

[3] Note that the $k^{-3/2}$ spectrum was also derived *analytically* by Kraichnan by using a technique called direct interaction approximation (DIA) first used in hydrodynamics (Kraichnan, 1958). It is, however, not asymptotically exact for HD because it fails invariance under random Galilean transformations, and for MHD because it does not include local anisotropy.

[4] Turbulence in liquid metals is also characterized by velocity spectra close to $k^{-5/3}$ over more than three orders of magnitude (Crémer and Alemany, 1981).

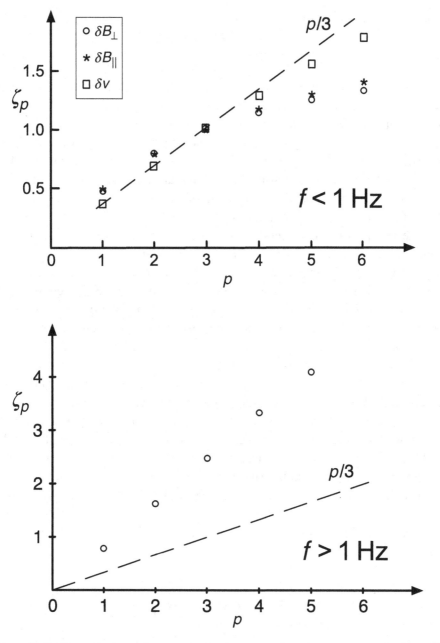

Figure 12.7 Top: velocity and magnetic field structure-function exponents measured in the solar wind at low frequencies ($f < 1$ Hz) – adapted from Carbone *et al.* (2004). Bottom: reproduction of similar measurements only for the magnetic field fluctuations at high frequencies ($f > 1$ Hz) – adapted from Kiyani *et al.* (2009). In both cases, the self-similar Kolmogorov law in $p/3$ is given as a dashed line.

intermittency is stronger for the magnetic field with a very significant deviation from the Kolmogorov self-similarity law in $p/3$. On the other hand, the situation is different for the magnetic fluctuations[5] at high frequencies (bottom): a self-similar law approximately in $5p/6$ appears, which is consistent with a magnetic spectrum in $-8/3$ (see Figure 12.1).

The major difference between the heliospheric low- and high-frequency plasma fluctuations is that, in the first case, the standard MHD seems to be a good model, whereas in the second it is necessary to add new effects to account for the ion inertia, for example the Hall effect. Kinetic effects such as Landau damping can also play a role, as can the anisotropy, which is important at all scales.

12.5   Magnetic Helicity and Inverse Cascade

Until now we have been interested mainly in energy and direct cascade processes, namely a transfer from large to small scales. In some situations, we may have the opposite phenomenon with a cascade from small to large scales: one speaks of an **inverse cascade**. In the two-dimensional hydrodynamic case it was Kraichnan[6] who suggested for the first time in 1967 (Kraichnan, 1967) the existence of an inverse cascade for the energy with a Kolmogorov spectrum in $-5/3$. This proposition has since been verified in laboratory experiments (Paret and Tabeling, 1998) and by direct numerical simulations. The two-dimensional approach is relevant when one wants to study, for example, the large-scale terrestrial atmospheric dynamics, because the vertical scale is significantly smaller than the horizontal one.

In the MHD case, it seems less relevant to focus on two-dimensional turbulence, on the one hand because plasmas are by nature three-dimensional and, on the other hand, because two-dimensional energy continues to cascade directly to small scales. However, an inverse cascade is possible in three-dimensional MHD for the third ideal invariant, namely the magnetic helicity. This analytical prediction made by Frisch *et al.* (1975) was verified numerically for the first time in 1976 (Pouquet *et al.*, 1976) using a spectral EDQNM model (see Section 13.2 and e.g. Grappin *et al.* (1982)). Figure 12.8 shows the principle of an inverse cascade: a stimulation of the three-dimensional MHD system at an intermediate

---

[5] At high frequencies ($f > 1$ Hz), only the magnetic field is accessible because the minimum integration time to build the velocity field from the distribution functions is of the order of a second.

[6] Robert Kraichnan (1928–2008) – the last postdoctoral student of Albert Einstein – received the Dirac ICTP medal in 2003 for his work on turbulence and specifically for his discovery of the inverse cascade in two-dimensional hydrodynamics. He shared this prize with Vladimir Zakharov who was distinguished for his work on weak wave turbulence and more particularly for having rigorously formalized that regime and derived the turbulence spectra as exact solutions of the asymptotic equations. Note that the notion of an inverse cascade was already present (but not totally understood) in the 1966 PhD thesis of Zakharov, who worked on gravity-wave turbulence.

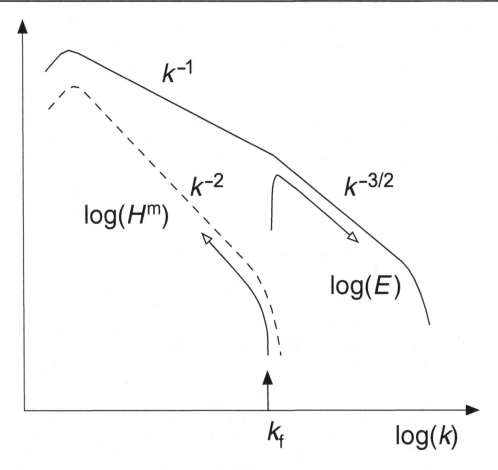

Figure 12.8 When a three-dimensional MHD system is excited at an intermediate scale k_f, an inverse cascade of magnetic helicity H^m can be triggered and a stationary spectrum in k^{-2} (dash) may be formed. With this cascade we can associate an energy spectrum in k^{-1} which is not the result of an energy cascade but just the signature of magnetic helicity at large scale. On the other hand, energy can cascade directly to small scales following, for example, a spectrum in $k^{-3/2}$ (solid).

scale k_f can generate both an inverse magnetic helicity cascade in k^{-2} and a direct energy cascade. Note that to get this dual cascade the stimulation must involve an injection of both magnetic helicity and energy fluxes.

To demonstrate the heuristic law $H^m(k) \sim k^{-2}$, we use the IK phenomenology and therefore we introduce a transfer time built on the eddy-turnover time ($\tau_{\text{eddy}} \sim \ell/b_\ell$; we assume equipartition between the kinetic and magnetic energies) and the Alfvén time ($\tau_A \sim \ell/b_0$). We have trivially

$$\varepsilon^H \sim \frac{h_\ell^m}{\tau_{tr}} \sim \frac{h_\ell^m}{\tau_{\text{eddy}}^2/\tau_A} \sim \frac{h_\ell^m b_\ell^2}{\ell b_0}. \tag{12.28}$$

Table 12.1 Cascade directions in three-dimensional isotropic turbulence. In Hall MHD, the generalized helicity may lead to a direct (inverse) cascade if e.g. the kinetic energy is much larger (smaller) than the magnetic energy.

	HD	MHD	Hall MHD	EMHD
Energy	Direct	Direct	Direct	Direct
Kinetic helicity	Direct			
Cross-helicity		Direct		
Magnetic helicity		Inverse	Inverse	Inverse
Generalized helicity			Direct/inverse	

By definition $h^{\mathrm{m}} \equiv \langle \mathbf{a} \cdot \mathbf{b} \rangle$ (for a velocity normalization), hence we have the dimensional relationship

$$h_\ell^{\mathrm{m}} \sim \ell b_\ell^2. \tag{12.29}$$

This gives (with $h_\ell^{\mathrm{m}} \sim H^{\mathrm{m}}(k)k$)

$$\varepsilon^{\mathrm{H}} \sim \frac{h_\ell^{\mathrm{m}2}}{\ell^2 b_0} \sim \frac{(H^{\mathrm{m}}(k)k)^2 k^2}{b_0}, \tag{12.30}$$

which leads to the **magnetic helicity spectrum**:

$$\boxed{H^{\mathrm{m}}(k) \sim \sqrt{\varepsilon^{\mathrm{H}} b_0}\, k^{-2},} \tag{12.31}$$

with[7] ε^{H} the mean rate of magnetic helicity transfer.[8] By dimensional analysis, it can be shown that this spectral law is compatible with an energy spectrum in k^{-1} (with $h_\ell^{\mathrm{m}} \sim \ell E_\ell$, which corresponds to the assumption of a maximal-helicity state[9]). In other words, if we have a source of energy and of helicity at an intermediate scale, a dual power law emerges for the energy spectrum: in $-3/2$ (or $-5/3$ according to the phenomenology) at small scales and in -1 at large scales. The nature of these laws is different: the first is the result of a direct energy cascade while the second is the manifestation of the inverse magnetic helicity cascade. Note that the -2 scaling derived phenomenologically does not exclude the possibility of getting other solutions. This is actually an open question which is under investigation.

A summary of the inviscid/ideal invariants and their cascade directions in hydrodynamics, MHD, Hall MHD, and electron MHD is given in Table 12.1.

---

[7] The Greek letter η is often used to define the mean rate of helicity transfer because originally the Roman letter H comes from η. For an obvious reason, in MHD we prefer to use the symbol ε^{H}.

[8] Curiously, it can be noted that the Kolmogorov phenomenology ($\tau_{\mathrm{tr}} = \tau_{\mathrm{eddy}}$) gives the same scaling law: $H^{\mathrm{m}}(k) \sim (\varepsilon^{\mathrm{H}})^{2/3} k^{-2}$.

[9] The energy and magnetic helicity spectra satisfy the Schwarz inequality $E(\mathbf{k}) \leq k|H^{\mathrm{m}}(\mathbf{k})|$. The equality between the two terms is reached in the maximal-helicity state, which is not obviously maintained over a wide range of scales. Therefore, we have to keep in mind that it is just a useful (but strong) assumption.

Note that the predictions concern only three-dimensional isotropic systems, in which in particular no external agent is present (such as a uniform magnetic field, rotation, stratification, etc).

12.6   The Critical Balance Conjecture

Most astrophysical plasmas evolve in media imbedded in a quasi-uniform magnetic field $\mathbf{b}_0$. An illustration is given by the heliospheric plasma, since on average the magnetic field forms a spiral (known as Parker's spiral) with an amplitude of the order of the fluctuations (moderate b_0). The Sun with its coronal loops is a second example; in this case, the mean magnetic field is significantly larger than the fluctuations (strong b_0).

Unlike in hydrodynamics, where a mean flow can be eliminated by a Galilean transformation, in MHD a mean magnetic field cannot be removed by such a transformation. Then, it is legitimate to ask whether MHD turbulence behaves in the same way in the parallel and transverse directions to the mean magnetic field $\mathbf{b}_0$. All theoretical, numerical, and observational analyses show that this is not the case and that MHD turbulence becomes anisotropic[10] with a cascade that grows preferentially in the transverse direction to $\mathbf{b}_0$. It is therefore essential to make this distinction – even phenomenologically – in our analysis of MHD turbulence. In this section, we present a phenomenological model for a **moderate** b_0 called critical balance (Higdon, 1984; Goldreich and Sridhar, 1995).

We shall consider the MHD equations for Alfvén wave-packets in which we forget the dissipative terms that have a negligible role in the inertial range (a balanced turbulence will also be assumed, i.e. $z^+ \sim z^- \sim z$):

$$\frac{\partial \mathbf{z}^{\pm}}{\partial t} \mp \mathbf{b}_0 \cdot \nabla \mathbf{z}^{\pm} + \mathbf{z}^{\mp} \cdot \nabla \mathbf{z}^{\pm} = -\nabla P_* . \tag{12.32}$$

The original idea of the critical balance is that in the inertial range a balance between the linear term which characterizes the transport of a wave-packet and the non-linear term which corresponds to its deformation appears naturally (regardless of the scale ℓ considered). In other words, this means that the Alfvén time τ_A balances the eddy-turnover time τ_{eddy} at **each scale**:

$$\boxed{\tau_A \sim \tau_{\text{eddy}}} . \tag{12.33}$$

These time scales can be estimated in the presence of anisotropy; we obtain

$$\tau_A \sim \frac{1}{k_\parallel b_0} \quad \text{and} \quad \tau_{\text{eddy}} \sim \frac{1}{k_\perp z_\ell} . \tag{12.34}$$

---

[10] As shown in Section 13.2.1, a simple explanation for the anisotropy can be given from the resonance condition that appears in the weak wave turbulence regime (see also e.g. Oughton *et al.* (1994)).

The first estimate is trivial (see Figure 12.6); for the second we implicitly assume that a (local) uniform magnetic field leads to a cascade mainly in the perpendicular direction. In this case, it is expected that energy will fill eventually higher perpendicular wavenumbers than parallel wavenumbers. It is then reasonable to assume that far enough into the inertial range the inequality $k_\perp \gg k_\parallel$ is satisfied. We will return to this point later in the framework of weak Alfvén wave turbulence. From these new definitions and the relationship (12.33), we obtain

$$z_\ell \sim b_0 \frac{k_\parallel}{k_\perp} . \tag{12.35}$$

Insofar as there is a balance between the only two available time scales, the phenomenology reduces to that of Kolmogorov with $\tau_{tr} \sim \tau_{eddy} \sim \tau_A$. This relation can also be interpreted in terms of collisions between wave-packets: a deformation of order one of the Alfvén wave-packet appears after only one collision. Then, we obtain trivially

$$\varepsilon \sim \frac{z_\ell^2}{\tau_{tr}} \sim k_\perp z_\ell^3 . \tag{12.36}$$

The combination of relations (12.35) and (12.36) ultimately gives the scaling relation for the **critical balance**:

$$k_\parallel \sim \left(\frac{\varepsilon^{1/3}}{b_0} \right) k_\perp^{2/3} \sim \ell_0^{-1/3} k_\perp^{2/3} . \tag{12.37}$$

This relationship expresses the fact that the MHD cascade develops preferentially – but not exclusively – in the transverse direction to the magnetic field $\mathbf{b}_0$ and that this turbulence becomes increasingly anisotropic at small scales as shown in Figure 12.9. In practice, the difficulty is in determining the parallel direction, which may vary slightly according to the scale that we choose. It can be shown by direct numerical simulations, and for a uniform magnetic field at least three to four times greater than the fluctuations, that this variation is negligible. On the other hand, even if a quasi-uniform magnetic field is not present at large scales, some studies (Müller, 2009) show that a local definition of the mean magnetic field is sufficient to define the parallel and transverse directions and then to verify the critical balance law (12.37). From this point of view, one recovers a central idea of the IK phenomenology.

The critical balance relation (12.37) is nowadays fairly well verified, in particular, by direct numerical simulations (see e.g. Cho and Vishniac (2000)). From the observational point of view, its verification is more difficult for technical reasons: it requires one to measure the plasma in several directions at the same time, which

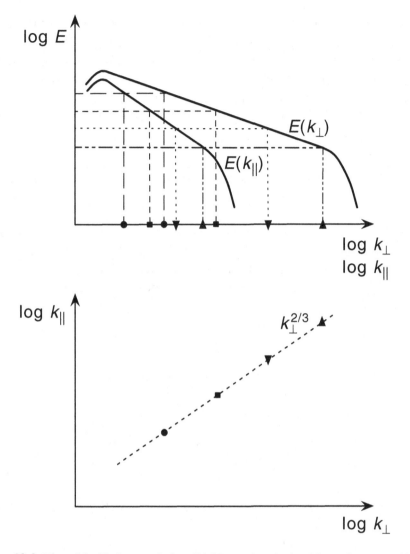

Figure 12.9 The critical balance relation (12.37) can be obtained from the perpendicular and parallel energy spectra (top), which demands in particular that we define properly (i.e. locally) the parallel direction. A given value of the energy corresponds to two points (symbols: points, squares, and triangles). The critical balance law emerges when these points are plotted (bottom).

is possible only if there are several satellites (see, however, Luo and Wu (2010)). The associated anisotropic spectrum,[11]

$$E(k_\perp) \sim k_\perp^{-5/3}, \tag{12.38}$$

---

[11] The value of the (Kolmogorov) constant in front of the right-hand-side term can be evaluated numerically. Its value is approximately 3.3 (Beresnyak, 2011).

is less difficult to verify since we can concentrate on a transverse direction. In practice, it appears that the solar wind satisfies this spectrum relatively well for the magnetic field but not for the velocity field, which follows a $-3/2$ scaling (Podesta *et al.*, 2007). Note that the critical balance solution can also be written in terms of a two-dimensional axisymmetric spectrum $E(k_\perp, k_\parallel) \sim k_\perp^{-5/3} k_\parallel^{-1}$, where by definition $\iint E(k_\perp, k_\parallel) dk_\perp dk_\parallel$ gives the total energy of the system. Then, it is straightforward to show that we also have $E(k_\parallel) \sim k_\parallel^{-2}$.

It is important to emphasize that even though the critical balance seems to surpass the isotropic IK approach it remains a phenomenology (this is not the case for the exact four-thirds law or the weak turbulence theory – see the next section) which is not able to explain the energy spectrum in $k_\perp^{-3/2}$ that is generally observed in three-dimensional direct numerical simulations when a moderate uniform magnetic field is present. To overcome this difficulty an alternative phenomenological model called dynamic alignment was proposed (Boldyrev, 2006). In this model the critical balance assumption is always used but the expression for the eddy-turnover time is modified in order to take into account the spontaneous alignment (to within an angle) of the **u** and **b** fields in the plane perpendicular to the field on approaching small scales. Then, a $k_\perp^{-3/2}$ spectrum may be predicted as well as the wavenumber dependence of the angle between **u** and **b**, though it is not obvious how one could verify this dependence numerically. Note finally that attempts are also being made to introduce a scaling relation like (12.37) at the level of the relationship (12.10) in order to obtain an axisymmetric Kolmogorov vectorial relation (Galtier, 2012).

12.7 Phenomenology of Weak Alfvén Wave Turbulence

We have seen that the IK phenomenology leads to the isotropic $k^{-3/2}$ spectrum. In this approach, the existence of a local magnetic field is assumed to justify the concept of an Alfvén wave-packet. As we have pointed out, the presence of a mean magnetic field is a source of anisotropy which comes into contradiction with the hypothesis of an isotropic IK spectrum. This contradiction in fact finds a solution in the particular regime of weak Alfvén wave turbulence where a strong external magnetic field is present. (This regime is opposed to the previous one, which is often called *strong* turbulence.) In the context of weak turbulence, it is possible to obtain analytically – by a rigorous perturbative approach – the dynamic equations of the system and to extract the stationary exact solutions by a conformal transformation (called Zakharov's transformation; see Nazarenko (2011)). Fortunately, this very heavy analytical demonstration[12] published in 2000 (Galtier *et al.*, 2000) can be explained with the IK phenomenology in which

---

[12] We give a partial demonstration of the theory of weak Alfvén wave turbulence in Chapter 13.

anisotropy is included through the characteristic time (12.34). In the simple case where the Alfvén waves fluxes are the same in both directions of propagation $(z^+ \sim z^- \sim z)$, we have the transfer time

$$\tau_{\text{tr}} \sim \frac{\tau_{\text{eddy}}^2}{\tau_A} \sim \frac{(\ell_\perp/z_\ell)^2}{\ell_\parallel/b_0} \sim \frac{k_\parallel b_0}{k_\perp^2 z_\ell^2}. \tag{12.39}$$

We thus obtain

$$\varepsilon \sim \frac{z_\ell^2}{\tau_{\text{tr}}} \sim \frac{k_\perp^2 z_\ell^4}{k_\parallel b_0} \sim \frac{k_\perp^2 (E(k_\perp, k_\parallel) k_\perp k_\parallel)^2}{k_\parallel b_0} \sim \frac{k_\perp^4 k_\parallel E^2(k_\perp, k_\parallel)}{b_0}, \tag{12.40}$$

and hence the anisotropic two-dimensional (axisymmetric) spectrum:

$$\boxed{E(k_\perp, k_\parallel) \sim \sqrt{\varepsilon b_0}\, k_\perp^{-2} k_\parallel^{-1/2}.} \tag{12.41}$$

This anisotropic MHD spectrum arises from the cumulative effect of the collisions of Alfvén wave-packets propagating in opposite directions. The phenomenology is, however, limited because the $k_\parallel$ dependence is only apparent: indeed, it can be shown rigorously that the non-linear transfer – the cascade – is completely frozen in the external magnetic field $\mathbf{b}_0$ direction.[13] Thus, the one and only relevant scaling is $k_\perp^{-2}$.

The $k_\perp^{-2}$ spectrum predicted analytically as an exact solution of the asymptotic equations was first reproduced by the simulation of the weak wave turbulence equations (Galtier et al., 2000), but we had to wait until 2015 to find the first strong evidence of this regime in a direct numerical simulation (Meyrand et al., 2015) (for weaker pieces of evidence see Bigot et al. (2008b) and Perez and Boldyrev (2008).) This result is quite remarkable given the required numerical resources (the difficulty lies in the fact that the presence of a strong uniform magnetic field imposes severe numerical restrictions). This confirmation suggests that the magnetic loops in the solar corona, and more broadly in stellar coronas, are probably in a regime of (highly anisotropic) weak Alfvén wave turbulence (see Figure 12.10). The study of turbulence by in situ measurements in the lower part of the solar corona is very difficult; however, a first step in that direction should be taken around 2018 with the NASA Solar Probe+ which will be able to come within a dozen solar radii of the sun. At this distance, it is not impossible that the regime of weak Alfvén wave turbulence might be detected. In fact, a first observational signature of this regime was found in Jupiter's magnetosphere, where the mean magnetic field is more than ten times higher than the fluctuations. With the help of in situ measurements made by the Galileo/ESA probe, the authors of this work showed spectra compatible with $k_\perp^{-2}$ predictions (Saur et al., 2002).

---

[13] See Section 13.2.1, for a simple explanation based on the resonance condition.

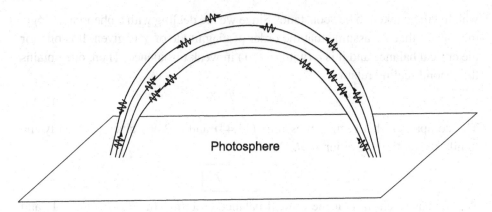

Figure 12.10 Schematic view of what could be the weak Alfvén wave turbulence regime on the Sun. Solar magnetic loops act as resonant cavities for Alfvén wave-packets produced under the photosphere and trapped in the loops. A wave-packet suffers numerous stochastic collisions during its propagation along the strong axial magnetic field. It is the cumulative effect of the collisions that leads to a significant distortion of the wave-packet: the deformation is associated with an energy cascade towards small scales.

12.8   (Grand) Unified Phenomenology

As we have seen, the critical balance spectrum and the weak turbulence spectrum can be derived by a different phenomenology where collisions between wave-packets are introduced. Whereas in the former case just one collision is able to produce a significant energy cascade, in the latter case one needs to consider the stochastic collisions of a large number of wave-packets. Although these approaches are different, we are going to show that a single phenomenology can describe both regimes.

Following the weak (balance) turbulence phenomenology, we write

$$\varepsilon \sim \frac{z_\ell^2}{\tau_{\rm tr}} \sim \frac{k_\perp^2 z_\ell^4}{k_\| b_0} \sim \frac{k_\perp^4 k_\| E^2(k_\perp, k_\|)}{b_0}. \tag{12.42}$$

Then, we introduce the two-dimensional axisymmetric spectrum:

$$E(k_\perp, k_\|) \sim k_\perp^{-\alpha} k_\|^{-\beta}, \tag{12.43}$$

where α and β are two unknown indices that we shall constrain. The introduction of (12.43) into expression (12.42) gives in particular the scaling relation

$$k_\| \sim k_\perp^{(4-2\alpha)/(2\beta-1)}. \tag{12.44}$$

We introduce now the time-scale ratio:

$$\chi \sim \frac{\tau_A}{\tau_{\rm eddy}} \sim \frac{k_\perp z_\ell}{k_\| b_0}, \tag{12.45}$$

which will be taken to be "constant." Since we are dealing with a phenomenology, this means that we assume that the order of magnitude of χ is given. It is one for the critical balance and it is ϵ (with $\epsilon \ll 1$) in weak turbulence. Then, one obtains the second scaling relation:

$$k_\parallel \sim k_\perp^{(3-\alpha)/(1+\beta)} . \tag{12.46}$$

The comparison between expressions (12.44) and (12.46) gives eventually the family of solutions (Galtier *et al.*, 2005)

$$\boxed{3\alpha + 2\beta = 7 .} \tag{12.47}$$

We can easily check that the critical balance spectrum ($\alpha = 5/3$, $\beta = 1$) and the weak turbulence spectrum ($\alpha = 2$, $\beta = 1/2$) satisfy well this relationship. In principle, this unified phenomenology can describe the transition between weak and strong turbulence if a power law is found, or other situations like $\alpha = 7/3$ and $\beta = 0$ which may also correspond to an absence of cascade along the parallel direction (as shown in Section 13.2 it is also a possible solution of weak turbulence). Interestingly, the measures of the power-law index of the reduced magnetic spectrum in active regions of the solar photosphere give values that often lie between -2 and $-7/3$ (Abramenko, 2005), which could be interpreted as the signature of a weak anisotropic MHD turbulence regime.

We have seen that two regimes of turbulence can be encountered in MHD: weak and strong turbulence. In the presence of a uniform magnetic field, strong turbulence seems to be well described by the critical balance phenomenology (with possibly a refinement). When this regime is established the one and only limitation for the cascade is the small-scale decoupling between ions and electrons from which standard MHD becomes irrelevant. Then, we need to add a correction to the fluid MHD equations (see the next section). For weak turbulence the situation is different. Indeed, an evaluation of the time-scale ratio χ shows a dependence on $k_\perp$. More precisely, if we take the weak turbulence spectrum (12.41), we obtain

$$\chi \sim k_\perp^{1/2} , \tag{12.48}$$

which means that χ increases with the weak turbulence cascade. If χ is very small at the largest scale of the system, it will necessarily reach one at some small scales (large $k_\perp$) and then MHD turbulence will become strong. The transition between the weak and strong (critical balance) regimes (see Figure 12.11) is considered as one of the most important properties of MHD turbulence.

12.9   Hall MHD Turbulence

We conclude this chapter with Hall MHD turbulence, which has been the subject of numerous works since the beginning of the twenty-first century. One of the

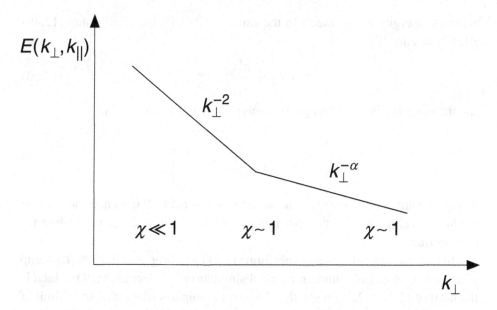

Figure 12.11 Schematic view of the transition between the weak and strong (balance) turbulence regimes which happens when χ becomes of order one. In this plot of the two-dimensional axisymmetric energy spectrum, $k_{\parallel}$ is fixed. In strong turbulence we may find e.g. $\alpha = 3/2$.

applications is the solar wind, in which the MHD inertial range is followed at higher frequencies by a second inertial range whose properties are partly compatible with those of Hall MHD.

12.9.1 The Four-Thirds Law and the Magnetic Spectrum

A first step in the study of a turbulent system is to get the equivalent of the Kolmogorov four-thirds law. This law was established for incompressible Hall MHD turbulence in 2008 (Galtier, 2008). The starting point of the analysis is the system (2.31)–(2.32). A similar (but slightly longer) calculation to the one developed in MHD (see expression (12.11)) leads to the four-thirds law for an isotropic turbulence:

$$-\frac{4}{3}\varepsilon^T \ell = \langle (\delta \mathbf{u} \cdot \delta \mathbf{u} + \delta \mathbf{b} \cdot \delta \mathbf{b})\delta u_\ell \rangle - 2\langle (\delta \mathbf{u} \cdot \delta \mathbf{b})\delta b_\ell \rangle + 4d_i \langle [(\mathbf{j} \times \mathbf{b}) \times \delta \mathbf{b}]_\ell \rangle,$$

(12.49)

where the last term on the right-hand side corresponds to the Hall effect. This exact law is very useful to better understand Hall MHD turbulence. Indeed, the Hall term introduces the magnetic field $\mathbf{b}$ but also the electric current $\mathbf{j}$, which dimensionally can be written as $j_\ell \sim b_\ell/\ell$. Thus, the Hall effect becomes relevant at small scales, i.e. when $\ell < d_i$ (equipartition between kinetic and magnetic energies is assumed). In the small-scale limit, the relationship (12.49) gives dimensionally

$$b_\ell^3 \sim \frac{\varepsilon^T}{d_i} \ell^2 , \tag{12.50}$$

and therefore the isotropic magnetic energy spectrum prediction is

$$E^b(k) \sim \left(\frac{\varepsilon^T}{d_i} \right)^{2/3} k^{-7/3} . \tag{12.51}$$

It is quite remarkable to see that the introduction of the Hall term does not change the linear scaling in ℓ for the four-thirds law whereas it modifies the scaling for the spectrum.

In fact, the magnetic energy spectrum (12.51) was proposed in 1996 (Biskamp *et al.*, 1996) to explain direct numerical simulations of electron MHD, which is the limit of (2.31)–(2.32) when the Hall term dominates (the small-scale limit of Hall MHD; see Section 2.4.2). In this case, the system reduces to (with $d_i = 1$)

$$\frac{\partial \mathbf{b}}{\partial t} = -\nabla \times ((\nabla \times \mathbf{b}) \times \mathbf{b}) + \eta \, \Delta \mathbf{b} , \tag{12.52}$$

which is significantly easier to simulate by computers than the complete Hall MHD equations. The Kolmogorov phenomenology for electron MHD can be written as

$$\varepsilon_\ell^T \sim \frac{b_\ell^2}{\tau_{tr}} \sim \frac{b_\ell^3}{\ell^2} , \tag{12.53}$$

where the transfer time, $\tau_{tr} \sim \ell^2/b_\ell$, is directly estimated from Eq. (12.52). Expression (12.53) is compatible with the magnetic energy spectrum (12.51). The exact law (12.49) is the only known example in turbulence where two inertial ranges can be described in a single expression. This is possible thanks to the additional richness brought by the study of conducting fluids.

12.9.2 Helicity Wave Turbulence

Like incompressible MHD, the weak wave turbulence regime may be investigated in incompressible Hall MHD. A comprehensive study was made in 2006 (Galtier, 2006), however, the helicity character of Hall MHD waves makes the development heavier than in the MHD case. For that reason, we limit ourselves to giving the power-law form of the exact solution of the asymptotic equations, namely the total energy spectrum:

$$E^T(k_\perp, k_\parallel) \sim \sqrt{\varepsilon^T b_0} \, k_\perp^{-2} k_\parallel^{-1/2} (1 + k_\perp^2 d_i^2)^{-1/4} . \tag{12.54}$$

As in the case of strong turbulence (see the previous section), we find that the energy spectrum stiffens at small scales (for $k_\perp d_i > 1$) but only in the transverse direction to the uniform magnetic field $\mathbf{b}_0$. As it is essentially in this direction that the non-linear transfer takes place, this is a fundamental item of information on the weak wave turbulence dynamics: a net stiffening of the spectrum is expected, in particular for its magnetic part that can be measured in the solar wind. We see in Figure 12.1 that actually a spectral stiffening whose exponent is close enough to the prediction (12.54) is observed. One of the current challenges is the need to understand the nature of the turbulent magnetic fluctuations of the solar wind at high frequency: one of the possible scenarios – but this is still contentious – is the emergence of a turbulence regime of right-polarized dispersive waves, with the left-polarized waves being less relevant at small scales. This scenario is somewhat reinforced by the presence of a self-similar intermittent law at high frequencies (see Figure 12.7), which is a natural property of other helical weak wave turbulence systems.

Advanced MHD Turbulence

In its primitive form the Kolmogorov theory states that the four-fifths law can be generalized to higher-order structure functions according to relation (11.33) by assuming self-similarity. Experiments and numerical simulations clearly show a discrepancy from this prediction (see Figure 11.10): this is what is commonly called *intermittency*. Even if intermittency remains a still poorly understood property of turbulence because it still challenges any attempt at a rigorous analytical description from first principles (i.e. the Navier–Stokes equations), several models have been proposed to reproduce the statistical measurements, of which the simplest is probably the fractal model, also called the β model, which was introduced in 1978 (Frisch *et al.*, 1978). As we shall see, this model is based on the idea of a fractal (incompressible) cascade and is therefore inherently a self-similar model. However, because the structure-function exponents are not those predicted by the Kolmogorov theory, one speaks of intermittency and anomalous exponents. Refined models have also been proposed, and we will present in this chapter the two most famous models: the log-normal and log-Poisson models.

13.1 Intermittency

13.1.1 Fractals and Multi-fractals

The idea underlying the β fractal model is Richardson's cascade (Figure 11.7): at each step of the cascade the number of children vortices is chosen so that the volume (or the surface in the two-dimensional case) occupied by these eddies decreases by a factor β ($0 < \beta < 1$) compared with the volume (or surface) of the parent vortex. The β factor is a parameter less than one of the model to reflect the fact that the filling factor varies according to the scale considered: the smallest eddies occupy less space than the largest.

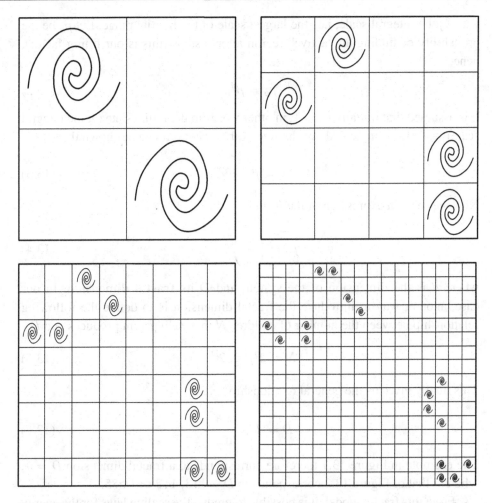

Figure 13.1 A fractal cascade in two dimensions for $\beta = 1/2$. At each stage of the cascade the basic scale is divided by two and the children vortices occupy only a fraction β of the surface of the parent vortex. At the integral scale (not shown) a single vortex occupies the entire available surface. The fractal dimension of this cascade is $D = 1$.

We define by ℓ_n the discrete scales of our system: the fractal cascade is characterized by jumps from the scale ℓ_n to the scale ℓ_{n+1}. We show an example of a fractal cascade in Figure 13.1: at each step of the cascade the elementary scale is divided by two.[1] So we have

$$\ell_n = \frac{\ell_0}{2^n}, \tag{13.1}$$

---

[1] Insofar as an inverse cascade occurs in two-dimensional hydrodynamic turbulence, Figure 13.1 should be seen as a simple illustration of the concept of a fractal cascade. In Figure 13.3, it is a three-dimensional fractal model that is considered.

with ℓ_0 the integral scale, i.e. the largest scale of the inertial range. Let p_n be the probability of finding an "active" region after n steps (this is our filling factor), hence

$$p_n = \beta^n . \tag{13.2}$$

It is assumed that initially $p_0 = 1$, in other words in the initial state (at the integral scale) we have a single eddy of the size of the system. It can be shown that

$$p_n = \beta^{\frac{\ln(\ell_n/\ell_0)}{\ln(1/2)}} . \tag{13.3}$$

Now we seek an expression of the form

$$p_n = \left(\frac{\ell_n}{\ell_0}\right)^{d-D} , \tag{13.4}$$

where d is the dimension of the system and D its **fractal dimension**. In our case, another way to introduce the fractal dimension is to define the following relationship between the number of children N that each parent produces and D:

$$N = 2^d \beta \equiv 2^D . \tag{13.5}$$

This finally gives us the general relationship

$$D = d + \frac{\ln \beta}{\ln 2} . \tag{13.6}$$

The example in Figure 13.1 therefore corresponds to a fractal dimension $D = 1$, whereas that of Figure 13.2 corresponds to $D = \ln 3/ \ln 2 \simeq 1.585$.

From this fractal model, it is possible to predict the scaling laws for the energy spectrum and more generally for the structure functions of order p. Eddies of size ℓ_n fill only a fraction p_n of the considered volume (we only study the three-dimensional case and take $d = 3$). Therefore, one can assume that the energy per unit mass associated with motions at the scale ℓ_n is such that

$$E_n = u_{\ell_n}^2 p_n = u_{\ell_n}^2 \left(\frac{\ell_n}{\ell_0}\right)^{3-D} . \tag{13.7}$$

Following the usual Kolmogorov phenomenology, one gets

$$\varepsilon \sim \frac{u_{\ell_n}^3}{\ell_n} p_n . \tag{13.8}$$

In particular, one has

$$\varepsilon \sim \frac{u_{\ell_0}^3}{\ell_0} , \tag{13.9}$$

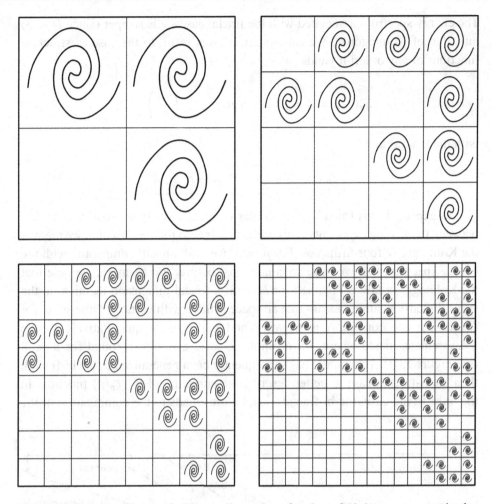

Figure 13.2 A fractal cascade in two dimensions for $\beta = 3/4$ (see comments in the caption of Figure 13.1). The fractal dimension of this cascade is $D \simeq 1.585$.

hence

$$u_{\ell_n} \sim u_{\ell_0} \left(\frac{\ell_n}{\ell_0} \right)^{(D-2)/3} . \tag{13.10}$$

Finally, we arrive at the following spectral prediction:

$$E_n \sim E(k)k \sim u_{\ell_0}^2 \left(\frac{\ell_n}{\ell_0} \right)^{(5-D)/3} , \tag{13.11}$$

or, in other words,

$$\boxed{E(k) \sim k^{-5/3-(3-D)/3} .} \tag{13.12}$$

The energy spectrum associated with the fractal cascade is steeper (since $D < 3$) than that of Kolmogorov. We can generalize this result to the case of structure functions of any order; it yields

$$S_p(\ell_n) = \langle (\delta u_{\ell_n})^p \rangle \sim p_n u_{\ell_n}^p \sim u_{\ell_0}^p \left(\frac{\ell_n}{\ell_0} \right)^{\zeta_p}, \tag{13.13}$$

with the **fractal law**

$$\boxed{\zeta_p = p/3 + (3 - D)(1 - p/3).} \tag{13.14}$$

As expected, this fractal – self-similar – model gives us a linear relationship between the scaling exponents. This correction is zero for $p = 3$ so that we recover the Kolmogorov four-fifths law. For $p = 2$, we find a result compatible with the energy spectrum that we have calculated just before. Furthermore, we note that the Kolmogorov law is obtained for $D = 3$: a fractal dimension identical to the space dimension simply means that all space is filled with vortices. Otherwise (for $D < 3$), the correction is negative and corresponds quantitatively to the measurements made where, for example, $D \simeq 2.8$ has been found for $p < 8$ (see Figure 13.3). This model has subsequently been generalized to the bi-fractal, then multi-fractal, case to better describe the curvature of the $\zeta_p(p)$ function. In the most trivial case, the bi-fractal model corresponds to the combination of the

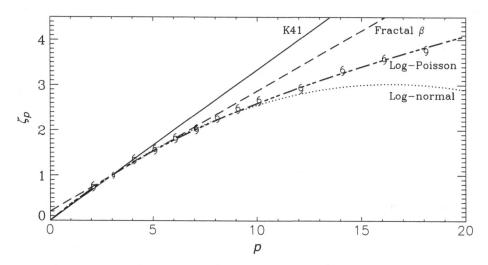

Figure 13.3 Reproduction of experimental measurements (symbols §) of the velocity structure-function exponents ζ_p. Four theoretical models of reference are plotted for comparison: the self-similar Kolmogorov model (K41), the three-dimensional β fractal model with $D = 2.8$ (dashes), the log-normal model with $\mu = 0.2$ (dots) and the log-Poisson model with $\beta = 2/3$ (dash–doted line).

Kolmogorov law for $p \leq 3$ and the fractal model for $p > 3$, a situation that we encounter in the Burgers turbulence.[2]

13.1.2 The Log-Normal Law

We have just seen that the β fractal model gives a different law from that of Kolmogorov in introducing the fundamental idea that intermittency is rooted in the non-uniform spatial distribution of turbulent structures. This property is observed very well in direct numerical simulations, with the appearance of clusters of vorticity filaments. As we see in Figure 13.3, however, the β fractal model becomes less relevant for high values of p where the data clearly show a curvature of the $\zeta_p(p)$ function. The log-normal model that we present in this section brings about this improvement by predicting a non-linear law in p. It was historically the first intermittency model: it was introduced by Kolmogorov (1962) in order to respond to a criticism made by Landau regarding the potential problems of the self-similar Kolmogorov theory.

The new idea introduced in the log-normal model is to rewrite the relationship (11.33) in the form

$$S_p(\ell) \equiv \langle (\delta u_\ell)^p \rangle = C_p \langle \varepsilon_\ell^{p/3} \rangle \ell^{p/3} , \tag{13.15}$$

with, by definition,

$$\varepsilon_\ell \equiv \frac{1}{(4/3)\pi \ell^3} \iiint_{|\mathbf{y}| \leq \ell} \varepsilon(\mathbf{x} + \mathbf{y}) dV , \tag{13.16}$$

which is the (local) dissipation of energy averaged over a ball of radius ℓ centered at $\mathbf{x}$. In this way, we will take into consideration Landau's comment about the possible fluctuations of the dissipation: for example $\langle \varepsilon_\ell \rangle^p$ will be even more different from $\langle \varepsilon_\ell^p \rangle$ the stronger the fluctuations of ε_ℓ, but these fluctuations tend to increase as the volume decreases since it is at small scales that dissipative structures are concentrated. (Note that we have the trivial relationship $\langle \varepsilon_\ell \rangle = \varepsilon$.) In this model, it is assumed that the probability density function of ε_ℓ follows a log-normal behavior, i.e. the probability density function of $\ln(\varepsilon_\ell)$ is a Gaussian of variance σ_ℓ centered on m_ℓ:

$$F'_{\ln(\varepsilon_\ell)} \equiv \frac{1}{\sqrt{2\pi \sigma_\ell^2}} \exp\left[-\frac{(\ln(\varepsilon_\ell/\varepsilon) - m_\ell)^2}{2\sigma_\ell^2} \right] . \tag{13.17}$$

---

[2] The Burgers equation is a prototype model of compressible turbulence. It can be written simply as $\partial_t u + u \, \partial_x u = \nu \, \partial_x^2 u$, where u is a scalar field which depends on x and t. The turbulence produced by this equation is bi-fractal with a transition at $p = 1$ (Frisch, 1995).

In passing, it may be noted that this of assumption Gaussianity cannot be made directly on the variable ε_ℓ, which is a strictly positive quantity. By a change of variables,[3] one obtains

$$F'_{\varepsilon_\ell} \equiv \frac{1}{\varepsilon_\ell} \frac{1}{\sqrt{2\pi\sigma_\ell^2}} \exp\left[-\frac{(\ln(\varepsilon_\ell/\varepsilon) + \sigma_\ell^2/2)^2}{2\sigma_\ell^2}\right], \qquad (13.18)$$

where we have used the relation $m_\ell = -\sigma_\ell^2/2$ which allows us to satisfy $\langle \varepsilon_\ell \rangle = \varepsilon$. We can deduce[4] that

$$\langle \varepsilon_\ell^n \rangle = \epsilon^n \exp\left(\frac{n\sigma_\ell^2}{2}(n-1)\right). \qquad (13.19)$$

Insofar as we seek power-law solutions, it is convenient to introduce the following form for the variance:

$$\sigma_\ell^2 \equiv \mu \ln\left(\frac{\ell_0}{\ell}\right), \qquad (13.20)$$

where μ is a constant that is supposed to be universal. We introduce (13.19) in (13.15), which eventually gives

$$S_p(\ell) = C_p(\varepsilon\ell_0)^{p/3} \left(\frac{\ell}{\ell_0}\right)^{\zeta_p}, \qquad (13.21)$$

with the **log-normal law**

$$\boxed{\zeta_p = \frac{p}{3} - \frac{\mu}{18}p(p-3).} \qquad (13.22)$$

The first intermittency measurements gave, with a relatively good accuracy, the anomalous exponents ζ_p for $p < 10$, and allow us to conclude that $\mu \simeq 0.2$. More recent measurements, however, show that for $p > 10$ a clear divergence appears (see Figure 13.3), invalidating the log-normal model. Despite these limitations, it is interesting to see what happens for $p = 2$; one gets $S_2(\ell) = C_2\varepsilon^{2/3}\ell^{2/3}(\ell/\ell_0)^{\mu/9}$, and hence the spectrum

$$\boxed{E(k) \sim \varepsilon^{2/3}k^{-5/3}(k\ell_0)^{-\mu/9}.} \qquad (13.23)$$

We see that the energy spectrum undergoes a slight correction of approximately -0.02 (with $\mu \simeq 0.2$) and thus becomes steeper. This is actually a trend observed in the experiens and numerical simulations. In conclusion, we can say that as a first approximation the log-normal distribution is relatively well observed, but that theoretically this model is not satisfactory: this observation motivated new work which culminated in the log-Poisson model.

---

[3] We recall the following mathematical property: the probability of a differential surface remains invariant under a change of variables i.e. $|F'_{\ln(\varepsilon_\ell)} d \ln(\varepsilon_\ell)| = |F'_{\varepsilon_\ell} d\varepsilon_\ell|$.

[4] We recall that: $\int_{-\infty}^{+\infty} \exp(-\alpha x^2 - \beta x)dx = \sqrt{\pi/\alpha}\exp[\beta^2/(4\alpha)]$, with $\alpha > 0$.

13.1.3 The Log-Poisson Law

The log-Poisson model of intermittency was proposed in 1994 (She and Leveque, 1994). It is currently the most used model because it reproduces very well the data to values up to $p \simeq 16$. It is based on the following three assumptions.

- Existence of a scaling law for structure functions:

$$S_p(\ell) \equiv \langle (\delta u_\ell)^p \rangle = C_p \langle \varepsilon_\ell^{p/3} \rangle \ell^{p/3} . \tag{13.24}$$

- Existence of a recurrence relation between the moments of the distribution of the local energy dissipation ε_ℓ:

$$\frac{\langle \varepsilon_\ell^{p+1} \rangle}{\varepsilon_\ell^\infty \langle \varepsilon_\ell^p \rangle} = A_p \left(\frac{\langle \varepsilon_\ell^p \rangle}{\varepsilon_\ell^\infty \langle \varepsilon_\ell^{p-1} \rangle} \right)^\beta , \quad 0 < \beta < 1, \tag{13.25}$$

where A_p are some constants and $\varepsilon_\ell^\infty \equiv \lim_{p \to \infty} \langle \varepsilon_\ell^{p+1} \rangle / \langle \varepsilon_\ell^p \rangle$ is a quantity that is mainly sensitive to the tail of the distribution of ε_ℓ. (We always have $\langle \varepsilon_\ell \rangle = \varepsilon$.)

- Existence of divergence scale dependence associated with the most intermittent dissipative structures:

$$\lim_{\ell \to 0} \varepsilon_\ell^\infty \sim \ell^{-2/3} . \tag{13.26}$$

Upon introducing the scaling relation $\langle \varepsilon_\ell^p \rangle \sim \ell^{\tau_p}$, expression (13.25) leads to the following recurrence relation:

$$\tau_{p+1} - (1 + \beta)\tau_p + \beta\tau_{p-1} + \frac{2}{3}(1 - \beta) = 0 . \tag{13.27}$$

By defining $\tau_p = -2p/3 + 2 + f_p$, we get a simple form of linear recursive progression of order two,

$$f_{p+1} - (1 + \beta)f_p + \beta f_{p-1} = 0 , \tag{13.28}$$

that we can solve easily; we find $f_p = \lambda + \mu\beta^p$. The coefficients λ and μ are determined from the initial conditions $\tau_0 = \tau_1 = 0$, hence

$$\tau_p = \frac{2}{3} \left(\frac{1 - \beta^p}{1 - \beta} - p \right) . \tag{13.29}$$

Finally, we obtain the following **log-Poisson law** (with by definition, $S_p(\ell) \sim \ell^{\zeta_p}$):

$$\boxed{\zeta_p = \frac{p}{3} + \frac{2}{3} \left(\frac{1 - \beta^{p/3}}{1 - \beta} - \frac{p}{3} \right) .} \tag{13.30}$$

We see that $\lim_{\beta \to 1} \zeta_p = p/3$: the parameter β measures the degree of intermittency in the sense that the smaller the value, the stronger the intermittency.

The best agreement with the data is obtained for $\beta = 2/3$ (see Figure 13.3). The log-Poisson law correctly predicts the anomalous exponents up to approximately $p = 16$. The latest measurements seem to show that beyond 16 a divergence appears, perhaps revealing the limits of the model. Note, however, that precautions have to be taken in the statistical analysis because the uncertainties increase with p since the statistical sample is finite. Probably 16 is already too high for a serious comparison with a model.

The log-Poisson model owes its name to the fact that the recurrence relation (13.25) can be interpreted as an underlying statistical property of the distribution ε_ℓ (Dubrulle, 1994). Let us rewrite this recurrence for the variable $\pi_\ell = \varepsilon_\ell / \varepsilon_\ell^\infty$; we get by a simple substitution:

$$\langle \pi_\ell^{p+1} \rangle = A_p \frac{\langle \pi_\ell^p \rangle^{\beta+1}}{\langle \pi_\ell^{p-1} \rangle^\beta} . \tag{13.31}$$

We show easily that

$$\langle \pi_\ell^p \rangle = A_{p-1} \left(A_{p-2} \frac{\langle \pi_\ell^{p-2} \rangle^{\beta+1}}{\langle \pi_\ell^{p-3} \rangle^\beta} \right)^{\beta+1} \frac{1}{\langle \pi_\ell^{p-2} \rangle^\beta} = A_{p-1} A_{p-2}^{\beta+1} \frac{\langle \pi_\ell^{p-2} \rangle^{\beta^2+\beta+1}}{\langle \pi_\ell^{p-3} \rangle^{\beta(\beta+1)}}$$

$$= ... = A_{p-1} A_{p-2}^{\beta+1} A_{p-3}^{\beta^2+\beta+1} ... A_0^{\beta^{p-1}+...+1} \langle \pi_\ell \rangle^{\beta^{p-1}+...+1}$$

$$= B_p \langle \pi_\ell \rangle^{\frac{1-\beta^p}{1-\beta}} , \tag{13.32}$$

where the coefficients B_p depend on A_q (with $0 \le q < p$). The distribution F'_{π_ℓ} corresponding to the moments (13.32) is a generalized Poisson distribution for the variable $Y = \ln \pi_\ell / \ln \beta$, namely:

$$F'_Y = \frac{e^{-\lambda} \lambda^Y}{Y!} , \tag{13.33}$$

where λ is the variance. With the change of variables, we get

$$\langle \pi_\ell^p \rangle = \int \pi_\ell^p \frac{e^{-\lambda} \lambda^Y}{Y!} e^{-Y \ln \beta} \, d\pi_\ell = \int e^{pY \ln \beta} \frac{e^{-\lambda} \lambda^Y}{Y!} \, dY = e^{-\lambda} \int \frac{(\lambda \beta^p)^Y}{Y!} \, dY$$

$$= e^{-\lambda(1-\beta^p)} . \tag{13.34}$$

The condition $p = 0$ fixes the variance:

$$\lambda = \frac{1}{\beta - 1} \ln \langle \pi_\ell \rangle . \tag{13.35}$$

Hence, eventually,

$$\langle \pi_\ell^p \rangle = e^{\frac{1-\beta^p}{1-\beta} \ln \langle \pi_\ell \rangle} = \langle \pi_\ell \rangle^{\frac{1-\beta^p}{1-\beta}} . \tag{13.36}$$

We can even show that the relationship (13.32) is still satisfied if the distribution F'_Y is the convolution of a Poisson distribution with any other distribution. In this case, a coefficient appears in the calculation and $B_p \ne 1$.

13.1.4 The Log-Poisson Law for MHD

A log-Poisson model for MHD was suggested for the first time in 1994 (Grauer *et al.*, 1994); it was modified later in 1997 (Horbury and Balogh, 1997) to better agree with the data. The difficulty was in deciding how to adapt the log-Poisson model to a flow where the dissipative structures are essentially current and vorticity sheets while in hydrodynamics these are vorticity filaments. In MHD, the first assumption of the log-Poisson model is written

$$S_p^\pm(\ell) \equiv \left\langle \left(\delta z_\ell^\pm\right)^p\right\rangle = C_p^\pm \left\langle \varepsilon_\ell^{\pm p/3}\right\rangle \ell^{p/3} \sim \ell^{\zeta_p^\pm}. \tag{13.37}$$

If it is assumed that the medium is balanced in the sense that $z^+ \sim z^- \sim z$, one can forget the dependence on $\pm$, which simplifies the calculation. In order to correctly reproduce the direct numerical simulations, it is necessary simply to change the parameter β which is set to $1/3$ in MHD (which corresponds in particular to sheets instead of filaments); hence the **MHD log-Poisson law**:

$$\boxed{\zeta_p^{\text{MHD}} = \frac{p}{9} + 1 - \left(\frac{1}{3}\right)^{p/3}.} \tag{13.38}$$

This law agrees well with numerical data for values $p \leq 8$ (beyond 8 the degree of accuracy becomes too weak). Finally, note a significant difference from the hydrodynamic case: in MHD we assume with expression (13.37) the existence of an exact relation for the variables $\left\langle \left(\delta z_\ell^\pm\right)^3\right\rangle$, which is not true, as can be seen from Eq. (12.17), where all the components of the Elsässer variables are present.

    A further difficulty appears in MHD because in most natural plasmas there exists a mean magnetic field $\mathbf{b}_0$. We have seen that this field generates anisotropy, it is therefore not surprising to find a trace of anisotropy in the anomalous exponents. To account for this, we write

$$S_p^\pm(\ell_\parallel) \equiv \left\langle \left(\delta z_{\ell_\parallel}^\pm\right)^p\right\rangle = C_{\parallel p}^\pm \left\langle \varepsilon_{\ell_\parallel}^{\pm p/3}\right\rangle \ell_\parallel^{p/3} \sim \ell_\parallel^{\zeta_{\parallel p}^\pm}, \tag{13.39}$$

$$S_p^\pm(\ell_\perp) \equiv \left\langle \left(\delta z_{\ell_\perp}^\pm\right)^p\right\rangle = C_{\perp p}^\pm \left\langle \varepsilon_{\ell_\perp}^{\pm p/3}\right\rangle \ell_\perp^{p/3} \sim \ell_\perp^{\zeta_{\perp p}^\pm}, \tag{13.40}$$

where the directions $\parallel$ and $\perp$ are parallel and perpendicular to $\mathbf{b}_0$, respectively. In Figure 13.4, we do indeed see that the anomalous exponents are very sensitive to the uniform field $\mathbf{b}_0$: different values are obtained for measurements made from the longitudinal ($S_p^\pm(\ell_\parallel)$) and transverse ($S_p^\pm(\ell_\perp)$) structure functions. These differences are even more marked the more intense the uniform field. In the case of the simulation with $b_0 = 0$, parallel and perpendicular directions are defined from the local average magnetic field. It is interesting to note that, even in this case where the system is statistically isotropic, (local) anisotropy can be detected.

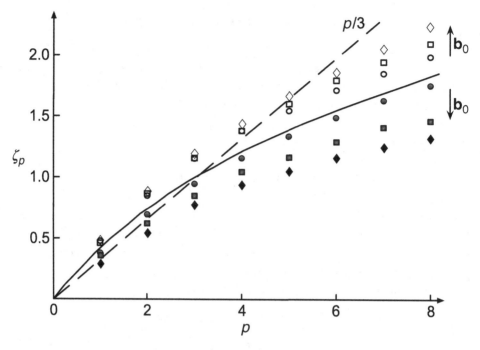

Figure 13.4 Anomalous exponents $\zeta_{\perp p}^{+}$ and $\zeta_{\parallel p}^{+}$ of structure functions computed from the Elsässer variables provided by direct numerical simulations. The log-Poisson model with $\beta = 1/3$ is plotted (solid line) as well as the self-similar Kolmogorov law (dashed line). Simulations are made in the presence of a uniform magnetic field b_0 along the z direction with $b_0 = 0$, 5, and 10 (circles, squares, and diamonds respectively) while fluctuations are of the order of unity. Empty and filled symbols correspond, respectively, to the longitudinal $S_p^{+}(\ell_{\parallel})$ and transverse $S_p^{+}(\ell_{\perp})$ structure functions (with, respectively, a separation ℓ parallel and transverse to the z direction). Adapted from Müller (2009).

13.2 Weak MHD Turbulence and the Closure Problem

We will now address the issue of a plasma subjected to a strong uniform magnetic field. The associated regime is called weak Alfvén wave turbulence and can be treated rigorously in spectral space. Weak wave turbulence corresponds to a regime where a large ensemble of waves will interact non-linearly. It is currently the subject of intensive research because modern experiments and simulations can reach this regime, and then confirm theoretical predictions and find new features (see e.g. Falcon *et al.* (2007), Meyrand and Galtier (2013), Shrira and Nazarenko (2013), Clark di Leoni *et al.* (2014), Yokoyama and Takaoka (2014), Aubourg and Mordant (2015), and Campagne *et al.* (2015)). In this section we present the asymptotic theory of weak MHD turbulence developed by Galtier *et al.* (2000). We will also discuss the closure issue, introduce the concept of triadic interaction for a whole class of turbulence problems including hydrodynamics, and demonstrate that the IK approach is relevant in this weak wave turbulence regime, provided that one incorporates anisotropy.

13.2.1   Triadic Interactions and Resonance

Let us write the inviscid and ideal MHD equations in Elsässer variables (to make the written expression more compact we use the polarization $s = \pm$):

$$\frac{\partial \mathbf{z}^s}{\partial t} - s\mathbf{b}_0 \cdot \nabla \mathbf{z}^s = -\mathbf{z}^{-s} \cdot \nabla \mathbf{z}^s - \nabla P_* , \tag{13.41}$$

where the parallel direction will be by definition the direction along the uniform field $\mathbf{b}_0$. We introduce the Fourier transform of the Elsässer fields (which is supposed to be defined):

$$z_j^s(\mathbf{x}, t) \equiv \iiint \hat{z}_j^s(\mathbf{k}, t) e^{i\mathbf{k}\cdot\mathbf{x}} \, d\mathbf{k} = \iiint \left[\epsilon a_j^s(\mathbf{k}, t) e^{is\omega_k t} \right] e^{i\mathbf{k}\cdot\mathbf{x}} \, d\mathbf{k} , \tag{13.42}$$

with $\omega_k \equiv k_\parallel b_0$. With the introduction of the variable a_j^s we are directly on the Alfvén wave-packet and we follow its (slow) evolution over time whose origin is purely non-linear. The parameter ϵ allows us to consider a field b_0 relatively strong compared with the fluctuations (in this case $\epsilon \ll 1$). Thus, after Fourier transform the MHD equations are

$$\frac{\partial a_j^s(\mathbf{k})}{\partial t} = -i\epsilon k_m P_{jn} \int_{\mathbf{R}^6} a_m^{-s}(\mathbf{q}) a_n^s(\mathbf{p}) e^{-is(\omega_k - \omega_p + \omega_q)t} \delta(\mathbf{k} - \mathbf{p} - \mathbf{q}) d\mathbf{p} \, d\mathbf{q} ,$$

$$\tag{13.43}$$

where $P_{jn}(k) \equiv \delta_{jn} - k_j k_n / k^2$ is the projection operator which maintains the fields transverse to $\mathbf{k}$ because of incompressibility. This expression is **exact** in the sense that it is strictly equivalent to the MHD Eqs. (13.41): no approximation has been carried out. Note that the derivation of expression (13.43) requires one first to write the total pressure as a function of the Elsässer variables. This is possible by applying the divergence operator to Eq. (13.41): $\Delta P_* = -\nabla \cdot (\mathbf{z}^{-s} \cdot \nabla \mathbf{z}^s)$.

Equation (13.43) shows that the evolution of an Alfvén wave-packet at a scale $\mathbf{k}$ is due to the non-linear interactions between two wave-packets of opposite polarity s. This evolution is slow since it occurs at the order ϵ. One can also see that the quadratic non-linearities are reflected in the spectral space by **triadic interactions** such as $\mathbf{k} = \mathbf{p} + \mathbf{q}$. In other words, all the non-linear interactions associated with a scale $\mathbf{k}$ correspond to an infinite number of triangles (Figure 13.5) whose form informs us on the local character of the interactions. It can be shown numerically that the local interactions (approximately equilateral triangles)

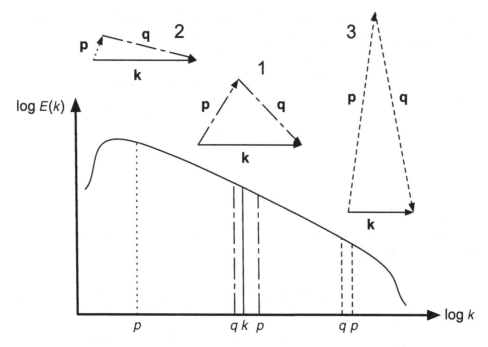

Figure 13.5 Examples of local (1) and non-local (2 and 3) triadic interactions with their positions on a spectrum.

contribute predominantly to the cascade. We thus recover the fundamental idea of a turbulent cascade operating scale-by-scale.[5]

Expression (13.43) reveals the presence of a complex exponential function, which gives in general a zero contribution at large times unless its exponent is null. Because we will study the wave-packet dynamics for an asymptotically long time, it is essential to check whether the three-wave **resonance condition** admits solutions; this condition is

$$\begin{cases} \omega_k = \omega_p - \omega_q, \\ \mathbf{k} = \mathbf{p} + \mathbf{q}. \end{cases} \tag{13.44}$$

It leads to (Shebalin *et al.*, 1983)

$$\boxed{q_\parallel = 0,} \tag{13.45}$$

and $\mathbf{k}_\perp = \mathbf{p}_\perp + \mathbf{q}_\perp$. This solution tells us that the deformation of an Alfvén wave-packet is directly related to its interaction with the two-dimensional part of the wave-packet propagating in the opposite direction. It also tells us that this interaction acts passively with the parallel component of the wavevector.

---

[5] The most recent numerical results show that isotropic MHD turbulence is a bit less local than hydrodynamic turbulence. However, it seems that the addition of a uniform magnetic field reinforces the local character (in the perpendicular direction) of MHD turbulence.

Therefore, only transverse non-linear transfers are expected in the long-time limit: weak Alfvén wave turbulence therefore proves to be infinitely anisotropic with a cascade occurring only in the transverse direction to the magnetic field $\mathbf{b}_0$.

13.2.2 IK Phenomenology Revisited

Now let us start again with the IK phenomenology and introduce a distinction between the parallel and transverse directions. We recall that $\tau_A \sim \ell_\parallel/b_0$ and $\tau_{\text{eddy}} \sim \ell_\perp/z_\ell$. We have seen that in a stochastic process of wave-packet collisions the transfer time is $\tau_{\text{tr}} \sim \tau_{\text{eddy}}^2/\tau_A$. As a result, we get (with $E \sim E^\pm$ for a balanced turbulence)

$$\varepsilon \sim \frac{z_\ell^2}{\tau_{\text{eddy}}^2/\tau_A} \sim \frac{z_\ell^4 \ell_\parallel}{\ell_\perp^2 b_0} \sim \frac{E^2(k_\perp, k_\parallel)k_\perp^4 k_\parallel}{b_0} . \tag{13.46}$$

Hence, the axisymmetric two-dimensional spectrum is

$$\boxed{E(k_\perp, k_\parallel) \sim \sqrt{\varepsilon b_0}\, k_\perp^{-2} k_\parallel^{-1/2} .} \tag{13.47}$$

Insofar as there is no parallel transfer, only the transverse scaling law is relevant. As we will see later, this phenomenological prediction in $k_\perp^{-2}$ is precisely the exact analytical solution obtained in weak Alfvén wave turbulence. This situation of balanced turbulence corresponds to zero cross-helicity. In the unbalanced case for which we have a non-zero cross-helicity ($E^+ \neq E^-$), we obtain a family of solutions such that

$$n_+ + n_- = -4, \tag{13.48}$$

with by definition $E^+ \sim k_\perp^{n_+}$ and $E^- \sim k_\perp^{n_-}$ (we have introduced in the phenomenology $\tau_{\text{eddy}}^\pm \sim \ell_\perp/z_\ell^\pm$ and $\varepsilon^\pm$). It is also the exact solution that we can derive from the weak wave turbulence equations.

13.2.3 Asymptotic Closure

We must now enter the heart of the matter and develop the formalism of weak wave turbulence. Instead of a rigorous and mathematically heavy demonstration, we choose to present the method with a general formalism.[6] We start from the characteristic equation

---

[6] The development exposed here is the classical one (Zakharov *et al.*, 1992; Benney and Newell, 1966) where in particular the Fourier space is supposed to be continuous. Recent analytical developments show that the introduction of a finite box with a discrete Fourier space will lead to new properties of weak turbulence (Nazarenko, 2011). This approach is particularly relevant for direct numerical simulation, where the Fourier space is always discretized.

$$\frac{\partial a_j(\mathbf{k})}{\partial t} = \epsilon \int_{\mathbf{R}^6} \mathcal{H}_{jmn}^{\mathbf{kpq}} a_m(\mathbf{p}) a_n(\mathbf{q}) e^{i\Omega_{k,pq}t} \delta_{k,pq} \, d\mathbf{p} \, d\mathbf{q}, \tag{13.49}$$

where $\delta_{k,pq} = \delta(\mathbf{k} - \mathbf{p} - \mathbf{q})$ and $\Omega_{k,pq} = \omega_k - \omega_p - \omega_q$; for simplicity, we ignore the time dependence of the variable $\mathbf{a}$. We note that

$$\mathcal{H}_{jmn}^{\mathbf{kpq}} = (\mathcal{H}_{jmn}^{-\mathbf{k}-\mathbf{p}-\mathbf{q}})^*, \tag{13.50}$$

$$\mathcal{H}_{jmn}^{\mathbf{kpq}} \text{ is symmetric in } (\mathbf{p}, \mathbf{q}) \text{ and } (m, n), \tag{13.51}$$

$$\mathcal{H}_{jmn}^{\mathbf{0pq}} = 0. \tag{13.52}$$

One defines the spectral energy tensor for a homogeneous turbulence:

$$\boxed{q_{jj'}(\mathbf{k}')\delta(\mathbf{k} + \mathbf{k}') \equiv \langle a_j(\mathbf{k}) a_{j'}(\mathbf{k}') \rangle.} \tag{13.53}$$

From (13.49), one finds

$$\frac{\partial q_{jj'} \delta(k + k')}{\partial t} = \left\langle a_{j'}(\mathbf{k}') \frac{\partial a_j(\mathbf{k})}{\partial t} \right\rangle + \left\langle a_j(\mathbf{k}) \frac{\partial a_{j'}(\mathbf{k}')}{\partial t} \right\rangle$$

$$= \varepsilon \int_{\mathbf{R}^6} \mathcal{H}_{jmn}^{\mathbf{kpq}} \langle a_m(\mathbf{p}) a_n(\mathbf{q}) a_{j'}(\mathbf{k}') \rangle e^{i\Omega_{k,pq}t} \delta_{k,pq} \, d\mathbf{p} \, d\mathbf{q}$$

$$+ \varepsilon \int_{\mathbf{R}^6} \mathcal{H}_{j'mn}^{\mathbf{k'pq}} \langle a_m(\mathbf{p}) a_n(\mathbf{q}) a_j(\mathbf{k}) \rangle e^{i\Omega_{k',pq}t} \delta_{k',pq} \, d\mathbf{p} \, d\mathbf{q}. \tag{13.54}$$

A hierarchy of equations appears clearly; for the third-order moment one has

$$\frac{\partial \langle a_j(\mathbf{k}) a_{j'}(\mathbf{k}') a_{j''}(\mathbf{k}'') \rangle}{\partial t}$$

$$= \varepsilon \int_{\mathbf{R}^6} \mathcal{H}_{jmn}^{\mathbf{kpq}} \langle a_m(\mathbf{p}) a_n(\mathbf{q}) a_{j'}(\mathbf{k}') a_{j''}(\mathbf{k}'') \rangle e^{i\Omega_{k,pq}t} \delta_{k,pq} \, d\mathbf{p} \, d\mathbf{q}$$

$$+ \varepsilon \int_{\mathbf{R}^6} \mathcal{H}_{j'mn}^{\mathbf{k'pq}} \langle a_m(\mathbf{p}) a_n(\mathbf{q}) a_j(\mathbf{k}) a_{j''}(\mathbf{k}'') \rangle e^{i\Omega_{k',pq}t} \delta_{k',pq} \, d\mathbf{p} \, d\mathbf{q}$$

$$+ \varepsilon \int_{\mathbf{R}^6} \mathcal{H}_{j''mn}^{\mathbf{k''pq}} \langle a_m(\mathbf{p}) a_n(\mathbf{q}) a_j(\mathbf{k}) a_{j'}(\mathbf{k}') \rangle e^{i\Omega_{k'',pq}t} \delta_{k'',pq} \, d\mathbf{p} \, d\mathbf{q}. \tag{13.55}$$

We can now write the fourth-order moment in terms of a sum of the fourth-order cumulant plus products of second-order ones, but a natural closure emerges for asymptotically long times (Benney and Newell, 1966). Within this limit several terms disappear, in particular, the fourth-order cumulant, which is not a resonant term. In other words, the non-linear regeneration of the third-order moments depends essentially on products of the second-order moment.[7] After a time integration, we get

---

[7] It is at this level of analysis that the closures used in turbulence are different. Research conducted mainly during the years 1960–1970 (Orszag, 1970) led in particular to a popular closure called EDQNM (eddy damped quasi-normal Markovian) which makes a link in an *ad hoc* way between the fourth-order cumulant and the third-order moment: the fourth-order cumulant plays the role of damping for the third-order moment without a memory effect. We may note that under this approximation no intermittency is present. See Section 12.5 for an application.

$$\langle a_j(\mathbf{k})a_{j'}(\mathbf{k}')a_{j''}(\mathbf{k}'')\rangle$$

$$= \varepsilon \int_{\mathbf{R}^6} \mathcal{H}^{\mathbf{kpq}}_{jmn}(\langle a_m(\mathbf{p})a_n(\mathbf{q})\rangle\langle a_{j'}(\mathbf{k}')a_{j''}(\mathbf{k}'')\rangle + \langle a_m(\mathbf{p})a_{j'}(\mathbf{k}')\rangle\langle a_n(\mathbf{q})a_{j''}(\mathbf{k}'')\rangle$$

$$+\langle a_m(\mathbf{p})a_{j''}(\mathbf{k}'')\rangle\langle a_n(\mathbf{q})a_{j'}(\mathbf{k}')\rangle)\Delta(\Omega_{k,pq})\delta_{k,pq}\, d\mathbf{p}\, d\mathbf{q}$$

$$+ \varepsilon \int_{\mathbf{R}^6} \mathcal{H}^{\mathbf{k'pq}}_{j'mn}(\langle a_m(\mathbf{p})a_n(\mathbf{q})\rangle\langle a_j(\mathbf{k})a_{j''}(\mathbf{k}'')\rangle + \langle a_m(\mathbf{p})a_j(\mathbf{k})\rangle\langle a_n(\mathbf{q})a_{j''}(\mathbf{k}'')\rangle$$

$$+ \langle a_m(\mathbf{p})a_{j''}(\mathbf{k}'')\rangle\langle a_n(\mathbf{q})a_j(\mathbf{k})\rangle)\Delta(\Omega_{k',pq})\delta_{k',pq}\, d\mathbf{p}\, d\mathbf{q}$$

$$+ \varepsilon \int_{\mathbf{R}^6} \mathcal{H}^{\mathbf{k''pq}}_{j''mn}(\langle a_m(\mathbf{p})a_n(\mathbf{q})\rangle\langle a_j(\mathbf{k})a_{j'}(\mathbf{k}')\rangle + \langle a_m(\mathbf{p})a_j(\mathbf{k})\rangle\langle a_n(\mathbf{q})a_{j'}(\mathbf{k}')\rangle$$

$$+ \langle a_m(\mathbf{p})a_{j'}(\mathbf{k}')\rangle\langle a_n(\mathbf{q})a_j(\mathbf{k})\rangle)\Delta(\Omega_{k'',pq})\delta_{k'',pq}\, d\mathbf{p}\, d\mathbf{q}, \tag{13.56}$$

where

$$\Delta(\Omega_{k,pq}) = \int_0^{t\gg 1/\omega} e^{i\Omega_{k,pq}t'}\, dt' = \frac{e^{i\Omega_{k,pq}t} - 1}{i\Omega_{k,pq}}. \tag{13.57}$$

After integration over the wavevectors $\mathbf{p}$ and $\mathbf{q}$, and simplification, one gets

$$\langle a_j(\mathbf{k})a_{j'}(\mathbf{k}')a_{j''}(\mathbf{k}'')\rangle = \varepsilon\Delta(\Omega_{kk'k''})\delta_{kk'k''}(\mathcal{H}^{\mathbf{k-k'-k''}}_{jmn}q_{mj'}(\mathbf{k}')q_{nj''}(\mathbf{k}'')$$

$$+ \mathcal{H}^{\mathbf{k-k''-k'}}_{jmn}q_{mj''}(\mathbf{k}'')q_{nj'}(\mathbf{k}')$$

$$+ \mathcal{H}^{\mathbf{k'-k-k''}}_{j'mn}q_{mj}(\mathbf{k})q_{nj''}(\mathbf{k}'') + \mathcal{H}^{\mathbf{k'-k''-k}}_{j'mn}q_{mj''}(\mathbf{k}'')q_{nj}(\mathbf{k})$$

$$+ \mathcal{H}^{\mathbf{k''-k-k'}}_{j''mn}q_{mj}(\mathbf{k})q_{nj'}(\mathbf{k}') + \mathcal{H}^{\mathbf{k''-k'-k}}_{j''mn}q_{mj'}(\mathbf{k}')q_{nj}(\mathbf{k})). \tag{13.58}$$

The symmetries (13.51) lead to

$$\langle a_j(\mathbf{k})a_{j'}(\mathbf{k}')a_{j''}(\mathbf{k}'')\rangle = 2\varepsilon\Delta(\Omega_{kk'k''})\delta_{kk'k''}(\mathcal{H}^{\mathbf{k-k'-k''}}_{jmn}q_{mj'}(\mathbf{k}')q_{nj''}(\mathbf{k}'')$$

$$+ \mathcal{H}^{\mathbf{k'-k-k''}}_{j'mn}q_{mj}(\mathbf{k})q_{nj''}(\mathbf{k}'')$$

$$+ \mathcal{H}^{\mathbf{k''-k-k'}}_{j''mn}q_{mj}(\mathbf{k})q_{nj'}(\mathbf{k}')). \tag{13.59}$$

The last expression can be inserted into (13.54). One takes the long time limit for which

$$\Delta(x) \to \pi\delta(x) + i\mathcal{P}(1/x), \tag{13.60}$$

with $\mathcal{P}$ the principal value of the integral. Finally, one obtains the **asymptotical** exact equations of weak wave turbulence:

$$\frac{\partial q_{jj'}(\mathbf{k})}{\partial t} = 4\pi\varepsilon^2 \int_{\mathbf{R}^6} \delta_{k,pq}\delta(\Omega_{k,pq})\mathcal{H}^{\mathbf{kpq}}_{jmn}$$

$$\times \left[\mathcal{H}^{\mathbf{p-q-k}}_{mrs}q_{rn}(\mathbf{q})q_{j's}(\mathbf{k}) + \mathcal{H}^{\mathbf{q-pk}}_{nrs}q_{rm}(\mathbf{p})q_{j's}(\mathbf{k}) \right.$$

$$\left. + \mathcal{H}^{\mathbf{-k-p-q}}_{j'rs}q_{rm}(\mathbf{p})q_{sn}(\mathbf{q}) \right] d\mathbf{p}\, d\mathbf{q}. \tag{13.61}$$

This expression is in principle valid for any situation where three-wave interactions are dominant. Only the form of $\mathcal{H}$ will be different. The equation for the energy spectrum is obtained by taking the trace of the spectral tensor $q_{ij}(\mathbf{k})$.

13.2.4 Exact Solutions in $k_\perp^{-2}$

In incompressible MHD, the asymptotic equation of weak (shear) Alfvén wave turbulence (within the limit $k_\perp \gg k_\parallel$) is (Galtier *et al.*, 2000)

$$
\frac{\partial E^\pm(k_\perp)}{\partial t} = \frac{\pi \epsilon^2}{b_0} \int_\Delta \cos^2 \phi \sin \theta \, \frac{k_\perp}{q_\perp} E^\mp(q_\perp)
$$
$$
\times \left[k_\perp E^\pm(p_\perp) - p_\perp E^\pm(k_\perp) \right] dp_\perp \, dq_\perp , \qquad (13.62)
$$

where Δ means that we must satisfy the triangular relationship $\mathbf{k}_\perp = \mathbf{p}_\perp + \mathbf{q}_\perp$ during the integration. It can be shown that $E^\pm(k_\perp, k_\parallel) = E^\pm(k_\perp)g^\pm(k_\parallel)$, with $g^\pm$ two arbitrary functions that express the lack of transfer along the parallel direction: its form depends essentially on the initial condition. In addition, ϕ is the angle associated with $(\mathbf{k}_\perp, \mathbf{p}_\perp)$ and θ the angle associated with $(\mathbf{k}_\perp, \mathbf{q}_\perp)$. Note that the nature of the term involving the wavevector $\mathbf{q}$ is different from the others since it corresponds to the two-dimensional mode (see relation (13.45)). The implicit assumption that we make below is that it behaves similarly to the three-dimensional modes.

The exact power-law solutions of these coupled integro-differential equations may be obtained by applying a conformal transformation – the Zakharov transformation (Zakharov, 1965; Nazarenko, 2011). In the spectral space we define

$$
\frac{\partial E^\pm(k_\perp)}{\partial t} = -\frac{\partial \varepsilon^\pm(k_\perp)}{\partial k_\perp}, \qquad (13.63)
$$

where $\varepsilon^\pm$ is the energy flux. This definition is in fact applicable to any incompressible turbulence. In the simplest case of a balanced turbulence (i.e. $E^\pm = E$; $\varepsilon^\pm = \varepsilon$), the constant (but non-zero) flux solution is[8]

$$
E(k_\perp) = C\sqrt{\varepsilon b_0}\, k_\perp^{-2} , \qquad (13.64)
$$

where C is the Kolmogorov constant whose analytical expression is not trivial; its numerical evaluation gives $C \simeq 0.585$. We can also prove rigorously that the energy flux is positive and thus the cascade is direct. The analytical solution is illustrated in Figure 13.6: one sees how the constant flux solution is formed with, in the first phase, propagation of the spectrum towards small scales and then the formation of the universal law in $k_\perp^{-2}$ over the entire inertial range. Figure 13.7

---

[8] The Zakharov transformation allows us to find another solution: the thermodynamic solution with zero energy flux. The associated spectrum in $E \sim k_\perp$ is, however, less relevant.

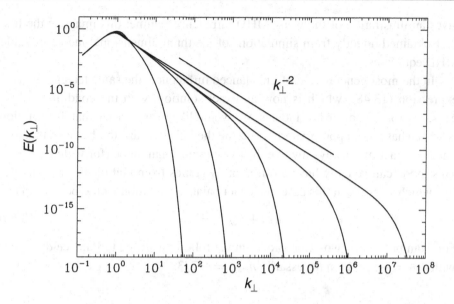

Figure 13.6 Numerical simulation of Eq. (13.62) in the case $E^{\pm} = E$. A spectrum in $E(k_{\perp}) \sim k_{\perp}^{-2}$ appears over more than six orders of magnitude. For reasons of numerical stability a dissipative term is added to the equation.

Figure 13.7 Three-dimensional direct numerical simulation of incompressible MHD. The magnetic field modulus is displayed in a section perpendicular to b_0. The weak turbulence regime appears as patchy structures instead of coherent structures as in strong turbulence. Courtesy of Dr. R. Meyrand (see also Meyrand *et al.* (2015)).

gives an illustration of the weak MHD turbulence regime: this image is the first ever obtained directly from simulations of the three-dimensional incompressible MHD equations.

In the most general case of unbalanced turbulence, the same technique gives expression (13.48), which is now an exact solution, with the condition $-3 < n_\pm < -1$ for the power-law indices. We would like to emphasize that this solution is somewhat more appropriate to describe the balance case too because the non-linear interactions always involve the two-dimensional mode (for which $q_\parallel = 0$). This mode can actually have a different dynamics from that of the wave modes (for which $k_\parallel > 0$). In this case, the exact balanced solution would be written as

$$n_{2D} + n_{wave} = -4 . \tag{13.65}$$

For example, if the two-dimensional mode follows a strong turbulence dynamics with $n_{2D} = -5/3$, then necessarily $n_{wave} = -7/3$.

Exercises for Part IV

IV.a Turbulent Diffusion Model

We propose to study a *ad hoc* turbulence model (Leith, 1969) based on the idea of the energy diffusion in spectral space. In practice, we consider the following equation:

$$\frac{\partial e(\mathbf{k})}{\partial t} = -\nabla \cdot \mathbf{F}(\mathbf{k}),$$

where $\mathbf{F}$ is the energy flux and $e(\mathbf{k})$ the energy spectrum.

(1) Explain why this equation is valid in the inertial range.
(2) To simplify the analysis we assume isotropy. Using spherical coordinates, rewrite the diffusion equation.
(3) We model the radial component of the energy flux in the following manner:

$$F_r = -D(k)\frac{\partial e(\mathbf{k})}{\partial k},$$

where D is a diffusion coefficient which remains to be determined. What is the dimension of D?
(4) We call τ the characteristic time of energy transfer. We focus firstly on the hydrodynamic case. Express τ with $e(\mathbf{k})$ and k.
(5) We introduce the reduced spectrum $E(k)$ such that $\int E(k)dk = \iiint e(\mathbf{k})d\mathbf{k}$. Show that the diffusion equation can be written in the form

$$\frac{\partial E(k)}{\partial t} = C\frac{\partial}{\partial k}\left(k^{11/2}\sqrt{E(k)}\frac{\partial(E(k)/k^2)}{\partial k}\right),$$

where C is a constant to be determined.
(6) We seek a particular solution of the above equation. We recall that the energy flux ε is defined by the relationship

$$\frac{\partial E(k)}{\partial t} = -\frac{\partial \varepsilon}{\partial k}.$$

What is the power-law solution corresponding to a constant (non-zero) energy flux?

(7) How this result is modified by the introduction of collisions of Alfvén wave-packets (in the isotropic case)?

IV.b Whistler Turbulence

We are interested in the purely whistler turbulence which is described by the electron MHD equation:

$$\partial_t \mathbf{b} = -d_i \nabla \times ((\nabla \times \mathbf{b}) \times \mathbf{b}) + \eta \, \Delta \mathbf{b}, \tag{13.66}$$

where $\mathbf{b}$ is the magnetic field normalized with respect to a velocity and d_i is the ion inertial length. The plasma is assumed to be incompressible.

(1) We assume isotropy. What is the characteristic energy-transfer time that emerges directly from the electron MHD equation?

(2) Deduce the form of the magnetic energy spectrum using a Kolmogorov phenomenology.

(3) We introduce a mean magnetic field $\mathbf{b}_0$ and therefore anisotropy into the system. Rewrite the electron MHD equation with this mean field. What is the expression for the whistler wave-packet collision time and how is the transfer time expressed?

(4) Deduce the shape of the axisymmetric spectrum of whistler turbulence.

Appendix I

Solutions to the Exercises

I.a Kinetic Helicity

We seek the equation of kinetic helicity conservation in incompressible MHD ($\rho = 1$). We have

$$\frac{\partial H^u}{\partial t} = \mathbf{u} \cdot \frac{\partial \mathbf{w}}{\partial t} + \mathbf{w} \cdot \frac{\partial \mathbf{u}}{\partial t},$$

with

$$\mathbf{w} \cdot \frac{\partial \mathbf{u}}{\partial t} = \mathbf{w} \cdot \left(-\nabla \left(\frac{u^2}{2} \right) + \mathbf{u} \times \mathbf{w} - \nabla P + \mathbf{j} \times \mathbf{b} + \nu \, \Delta \mathbf{u} \right),$$

$$\mathbf{u} \cdot \frac{\partial \mathbf{w}}{\partial t} = \mathbf{u} \cdot \nabla \times \left(-\nabla \left(\frac{u^2}{2} \right) + \mathbf{u} \times \mathbf{w} - \nabla P + \mathbf{j} \times \mathbf{b} + \nu \, \Delta \mathbf{u} \right).$$

One obtains

$$\begin{aligned}
\frac{\partial H^u}{\partial t} &= \mathbf{w} \cdot \left(-\nabla \left(\frac{u^2}{2} \right) - \nabla P + \mathbf{j} \times \mathbf{b} + \nu \, \Delta \mathbf{u} \right) \\
&\quad + \mathbf{u} \cdot (\nabla \times (\mathbf{u} \times \mathbf{w} + \mathbf{j} \times \mathbf{b} + \nu \, \Delta \mathbf{u})) \\
&= -\nabla \cdot \left(\frac{u^2}{2} \mathbf{w} + P\mathbf{w} \right) + \mathbf{w} \cdot (\mathbf{j} \times \mathbf{b}) + \nu \mathbf{w} \cdot \Delta \mathbf{u} \\
&\quad + \nabla \cdot ((\mathbf{u} \times \mathbf{w}) \times \mathbf{u}) + \nabla \cdot ((\mathbf{j} \times \mathbf{b}) \times \mathbf{u}) + (\mathbf{j} \times \mathbf{b}) \cdot \mathbf{w} + \nu \mathbf{u} \cdot \Delta \mathbf{w}.
\end{aligned}$$

Hence, the local form of the kinetic helicity conservation equation is

$$\begin{aligned}
\frac{\partial H^u}{\partial t} &= \nabla \cdot \left((\mathbf{u} \times \mathbf{w}) \times \mathbf{u} + (\mathbf{j} \times \mathbf{b}) \times \mathbf{u} - \frac{u^2}{2} \mathbf{w} - P\mathbf{w} \right) + 2(\mathbf{j} \times \mathbf{b}) \cdot \mathbf{w} \\
&\quad + \nu \mathbf{w} \cdot \Delta \mathbf{u} + \nu \mathbf{u} \cdot \Delta \mathbf{w}.
\end{aligned}$$

We see that the magnetic field provides a source term. The kinetic helicity is thus an inviscid invariant only in the hydrodynamic case. Finally, note that the kinetic helicity reflects the interlinkage of vorticity loops.

I.b Cross-Helicity and Topology

The proof is similar to that of the magnetic helicity. The cross-helicity of the magnetic tube T_1 (of infinitesimal section) is

$$\iiint_{T_1} H_1^c \, d\mathcal{V} = \iiint_{T_1} \mathbf{u} \cdot \mathbf{B} \, d\ell \, d\mathcal{S} = \iiint_{T_1} \mathbf{u} \cdot d\boldsymbol{\ell} \, B \, d\mathcal{S}$$

$$= \oint_{C_1} \mathbf{u} \cdot \left(\iint_{\delta s_1} B \, d\mathcal{S} \right) d\boldsymbol{\ell} = \oint_{C_1} \Phi_1 \mathbf{u} \cdot d\boldsymbol{\ell} = \Phi_1 \oint_{C_1} \mathbf{u} \cdot d\boldsymbol{\ell},$$

where C_1 is a closed curve representing the tube T_1, δs_1 is the section of the tube, and Φ_1 is the magnetic flux associated with it; this flux is constant along the tube and therefore can be put outside of the integral. We still have to evaluate the velocity circulation. By noticing that

$$\oint_{C_1} \mathbf{u} \cdot d\boldsymbol{\ell} = \iint_{S_1} \nabla \times \mathbf{u} \cdot \mathbf{n} \, d\mathcal{S} = \iint_{S_1} \mathbf{w} \cdot \mathbf{n} \, d\mathcal{S} = \Psi_2,$$

where S_1 is the surface defined by the tube T_1, $\mathbf{w}$ the vorticity and Ψ_2 the vorticity flux due to the vorticity tube T_2. One finds eventually

$$\iiint_{T_1} H_1^c \, d\mathcal{V} = \Phi_1 \Psi_2.$$

The cross-helicity may therefore account for the interlinkage of magnetic and vorticity tubes. Finally, note that a calculation of the cross-helicity from the vorticity tube provides a similar contribution. The total contribution is therefore

$$\iiint H_{\text{tot}}^c \, d\mathcal{V} = 2\Phi_1 \Psi_2.$$

The cross-helicity is thus a measure of the interlinkage of magnetic and vorticity tubes.

I.c Reduced MHD Approximation

For the resolution of the problem we will forget the viscous terms, which are not problematic. It is assumed that the plasma is strongly magnetized and the magnetic pressure dominates the kinetic pressure, which allows us to eliminate the pressure term. With the proposed notations, it yields

$$\partial_t \rho + \epsilon b_0 \, \nabla \cdot (\rho \mathbf{u}) = 0,$$

$$\partial_t \mathbf{u} + \epsilon b_0 \mathbf{u} \cdot \nabla \mathbf{u} = \frac{b_0}{\rho} \left[-\frac{1}{2} \nabla (2b_z + \epsilon b^2) + (\mathbf{e_z} + \epsilon \mathbf{b}) \cdot \nabla \mathbf{b} \right],$$

$$\partial_t \mathbf{b} + \epsilon b_0 \mathbf{u} \cdot \nabla \mathbf{b} = -b_0 (\mathbf{e_z} + \epsilon \mathbf{b})(\nabla \cdot \mathbf{u}) + b_0 (\mathbf{e_z} + \epsilon \mathbf{b}) \cdot \nabla \mathbf{u},$$

$$\nabla \cdot \mathbf{b} = 0.$$

One introduces the decompositions, $\mathbf{u} = \mathbf{u}_\perp + \mathbf{u}_z$, $\mathbf{b} = \mathbf{b}_\perp + \mathbf{b}_z$, and $\nabla = \nabla_\perp + \epsilon\, \partial_z$. Hence, after simplification,

$$\partial_t \rho + \epsilon b_0\, \nabla_\perp \cdot (\rho \mathbf{u}_\perp) + \epsilon^2 b_0\, \partial_z(\rho \mathbf{u}_z) = 0,$$

$$\partial_t \mathbf{u}_\perp + \epsilon b_0 \mathbf{u}_\perp \cdot \nabla_\perp \mathbf{u}_\perp + \epsilon^2 b_0 u_z \partial_z \mathbf{u}_\perp = -\frac{b_0}{2\rho}\, \nabla_\perp (2b_z + \epsilon b^2)$$

$$+ \frac{\epsilon b_0}{\rho}(\partial_z \mathbf{b}_\perp + \mathbf{b}_\perp \cdot \nabla_\perp \mathbf{b}_\perp + \epsilon b_z\, \partial_z \mathbf{b}_\perp),$$

$$\partial_t \mathbf{b}_\perp + \epsilon b_0 \mathbf{u}_\perp \cdot \nabla_\perp \mathbf{b}_\perp + \epsilon^2 b_0 u_z\, \partial_z \mathbf{b}_\perp = -b_0 \epsilon \mathbf{b}_\perp(\nabla_\perp \cdot \mathbf{u}_\perp + \epsilon\, \partial_z u_z) + b_0 \epsilon\, \partial_z \mathbf{u}_\perp$$

$$+ b_0 \epsilon (\mathbf{b}_\perp \cdot \nabla_\perp \mathbf{u}_\perp + \epsilon b_z\, \partial_z \mathbf{u}_\perp),$$

$$\partial_t u_z + \epsilon b_0 \mathbf{u}_\perp \cdot \nabla_\perp u_z + \epsilon^2 b_0 u_z\, \partial_z u_z = -\frac{\epsilon^2 b_0}{2\rho}\, \partial_z b^2$$

$$+ \frac{\epsilon b_0}{\rho}(\mathbf{b}_\perp \cdot \nabla_\perp b_z + \epsilon b_z\, \partial_z b_z),$$

$$\partial_t b_z + \epsilon b_0 \mathbf{u}_\perp \cdot \nabla_\perp b_z + \epsilon^2 b_0 u_z\, \partial_z b_z = -b_0\, \nabla_\perp \cdot \mathbf{u}_\perp$$

$$-b_0 \epsilon b_z(\nabla_\perp \cdot \mathbf{u}_\perp + \epsilon\, \partial_z u_z) + b_0 \epsilon (\mathbf{b}_\perp \cdot \nabla_\perp u_z + \epsilon b_z\, \partial_z u_z),$$

$$\nabla_\perp \cdot \mathbf{b}_\perp + \epsilon\, \partial_z b_z = 0.$$

At the main order, one gets

$$\partial_t \rho = -\epsilon b_0\, \nabla_\perp \cdot (\rho \mathbf{u}_\perp),$$

$$\partial_t \mathbf{u}_\perp + \frac{b_0}{\rho}\, \nabla_\perp b_z = -\epsilon b_0 \mathbf{u}_\perp \cdot \nabla_\perp \mathbf{u}_\perp - \frac{\epsilon b_0}{2\rho}\, \nabla_\perp b^2 + \frac{\epsilon b_0}{\rho}(\partial_z \mathbf{b}_\perp + \mathbf{b}_\perp \cdot \nabla_\perp \mathbf{b}_\perp),$$

$$\partial_t \mathbf{b}_\perp = -\epsilon b_0 \mathbf{u}_\perp \cdot \nabla_\perp \mathbf{b}_\perp - \epsilon b_0 \mathbf{b}_\perp(\nabla_\perp \cdot \mathbf{u}_\perp) + \epsilon b_0\, \partial_z \mathbf{u}_\perp + \epsilon b_0 \mathbf{b}_\perp \cdot \nabla_\perp \mathbf{u}_\perp,$$

$$\partial_t u_z = -\epsilon b_0 \mathbf{u}_\perp \cdot \nabla_\perp u_z + \frac{\epsilon b_0}{\rho}\mathbf{b}_\perp \cdot \nabla_\perp b_z,$$

$$\partial_t b_z + b_0 \nabla_\perp \cdot \mathbf{u}_\perp = -\epsilon b_0 \mathbf{u}_\perp \cdot \nabla_\perp b_z - \epsilon b_0 b_z(\nabla_\perp \cdot \mathbf{u}_\perp) + \epsilon b_0 \mathbf{b}_\perp \cdot \nabla_\perp u_z.$$

Since the density is related to the pressure by a polytropic relationship and this pressure is negligible compared with the magnetic pressure, we see that the density cannot vary at the order ϵ. Therefore $\nabla_\perp \cdot \mathbf{u}_\perp$ can be neglected. Under these conditions, the master dynamic equations are given by the perpendicular components (zero parallel components will remain unchanged)

$$\partial_t \mathbf{u}_\perp = -\epsilon b_0 \mathbf{u}_\perp \cdot \nabla_\perp \mathbf{u}_\perp - \epsilon b_0\, \nabla_\perp\left(\frac{b_\perp^2}{2}\right) + \epsilon b_0(\partial_z \mathbf{b}_\perp + \mathbf{b}_\perp \cdot \nabla_\perp \mathbf{b}_\perp),$$

$$\partial_t \mathbf{b}_\perp = -\epsilon b_0 \mathbf{u}_\perp \cdot \nabla_\perp \mathbf{b}_\perp + \epsilon b_0 \, \partial_z \mathbf{u}_\perp + \epsilon b_0 \mathbf{b}_\perp \cdot \nabla_\perp \mathbf{u}_\perp \, .$$

After a time renormalization $t \to \epsilon b_0 t$, we recover the proposed equations. Finally, note that a more complete demonstration can be achieved by making a multi-scale development in space and time (Zank and Matthaeus, 1992).

I.d Generalized Helicity

The demonstration is based on relations (3.37) and (3.43), as well as the relation obtained in the first exercise on the kinetic helicity. It remains to evaluate the contribution of $d_i \, \partial_t (\mathbf{a} \cdot (\nabla \times \mathbf{u}))$. It is easily shown that in the inviscid and ideal case all terms can be written in a divergence form.

II.a g and p Modes

If one disrupts a compressible medium subjected to gravity, it satisfies the following linear equations:

$$\omega P_1 = \rho_0 c_S^2 \mathbf{k} \cdot \mathbf{u}_1 \, ,$$

$$\omega \rho_1 = \rho_0 \mathbf{k} \cdot \mathbf{u}_1 \, ,$$

$$\omega \mathbf{u}_1 = \mathbf{k} \left(\frac{P_1}{\rho_0} \right) - \mathbf{k} \psi_1 \, ,$$

with $-k^2 \psi_1 = 4\pi \mathcal{G} \rho_1$. After rearrangement of the equations, we obtain

$$\omega^2 \mathbf{u}_1 = \left(c_S^2 - \frac{4\pi \mathcal{G} \rho_0}{k^2} \right) (\mathbf{k} \cdot \mathbf{u}_1) \mathbf{k} \, .$$

The associated waves are longitudinal. The angular-frequency solution to the problem is

$$\omega^2 = \left(c_S^2 - \frac{4\pi \mathcal{G} \rho_0}{k^2} \right) k^2 \, .$$

One speaks of a pressure wave – or p mode – when $c_S^2 > 4\pi \mathcal{G} \rho_0 / k^2$ and of a gravity wave – or g mode – when $c_S^2 < 4\pi \mathcal{G} \rho_0 / k^2$. In the second case the wave is unstable. In general, the instability tends to appear at the largest scales, i.e. at small values of k.

II.b Magnetostrophic Waves

(1) To find the linear solutions of incompressible rotating MHD, we introduce

$$\mathbf{b}(\mathbf{x}) \to \epsilon \mathbf{b}(\mathbf{x}) \, , \quad \mathbf{u}(\mathbf{x}) \to \epsilon \mathbf{u}(\mathbf{x}) \, ,$$

with ϵ a small parameter ($0 < \epsilon \ll 1$) and $\mathbf{x}$ a three-dimensional displacement vector; then we obtain the following inviscid and ideal equations in Fourier space:

$$\partial_t \mathbf{w_k} - 2ik_\| \Omega_0 \mathbf{u_k} - ik_\| b_0 \mathbf{j_k} = \epsilon \{\mathbf{w} \cdot \nabla \mathbf{u} - \mathbf{u} \cdot \nabla \mathbf{w} + \mathbf{b} \cdot \nabla \mathbf{j} - \mathbf{j} \cdot \nabla \mathbf{b}\}_\mathbf{k},$$

$$\partial_t \mathbf{b_k} - ik_\| b_0 \mathbf{u_k} = \epsilon \{\mathbf{b} \cdot \nabla \mathbf{u} - \mathbf{u} \cdot \nabla \mathbf{b}\}_\mathbf{k},$$

$$\mathbf{k} \cdot \mathbf{u_k} = 0,$$

$$\mathbf{k} \cdot \mathbf{b_k} = 0,$$

where the wavevector $\mathbf{k} = k\mathbf{e_k} = k_\perp + k_\| \mathbf{e}_\| $ ($|\mathbf{e_k}| = 1$). The index $\mathbf{k}$ denotes the Fourier transform, defined by the relation

$$\mathbf{u}(\mathbf{x}) \equiv \int \mathbf{u}(\mathbf{k}) e^{i\mathbf{k} \cdot \mathbf{x}} \, d\mathbf{k},$$

where $\mathbf{u}(\mathbf{k}) = \mathbf{u_k} = \tilde{\mathbf{u}}_\mathbf{k} e^{-i\omega t}$ (the same notation is used for the other fields). The linear dispersion relation ($\epsilon = 0$) reads

$$\omega^2 + \left(\frac{2\Omega_0 k_\|}{\Lambda k} \right) \omega - k_\|^2 b_0^2 = 0,$$

with

$$\begin{Bmatrix} \tilde{\mathbf{u}}_\mathbf{k} \\ \tilde{\mathbf{b}}_\mathbf{k} \end{Bmatrix} = \Lambda i \mathbf{e_k} \times \begin{Bmatrix} \tilde{\mathbf{u}}_\mathbf{k} \\ \tilde{\mathbf{b}}_\mathbf{k} \end{Bmatrix}.$$

We obtain the general solution (Galtier, 2014)

$$\omega \equiv \omega_\Lambda^s = \frac{sk_\| \Omega_0}{k} \left(-s\Lambda + \sqrt{1 + k^2 d^2} \right),$$

where the value ($\pm$) of s defines the directional wave polarity such that we always have $sk_\| \geq 0$; then ω_Λ^s is a positive definite angular frequency. Note the introduction of the magneto-inertial length $d \equiv b_0 / \Omega_0$, which can be seen as a measure of the relative importance of the Lorentz–Laplace and Coriolis forces.

(2) The plot of the dispersion relation for rotating MHD is given in Figure A1.1. We use the normalization angular frequency $X \equiv k\omega_\Lambda^s / (sk_\| \Omega_0)$ and the normalized wavenumber kd.

(3) The wave polarization Λ tells us whether the wave is right- ($\Lambda = s$) or left- ($\Lambda = -s$) circularly polarized. In the first case, we are dealing with the magnetostrophic branch, whereas in the latter case we have the inertial branch. We see that the transverse circularly polarized waves are dispersive and that we recover the two well-known limits, i.e. the pure inertial waves ($\omega_{-s}^s = 2s\Omega_0 k_\| / k \equiv \omega_I$) in the large-scale limit ($kd \to 0$) and the standard Alfvén waves ($\omega = sk_\| b_0 \equiv \omega_A$) in the small-scale limit ($kd \to +\infty$). For the pure

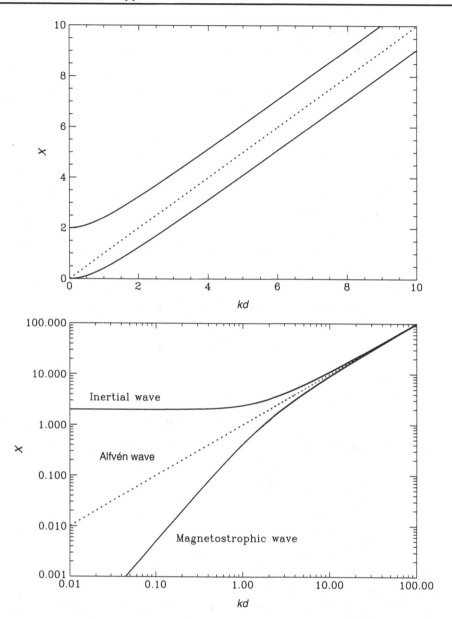

Figure AI.1 The dispersion relation for rotating MHD permeated by a background magnetic field in linear (top) and logarithmic (bottom) coordinates, with $X \equiv k\omega_A^s/(sk_{\parallel}\Omega_0)$. The upper and lower branches correspond, respectively, to left- and right-handed polarized waves. The Alfvén wave dispersion relation is also given (dotted line).

magnetostrophic waves we find the angular frequency $\omega_s^s = sk_{\parallel}kdb_0/2 = \omega_A^2/\omega_I \equiv \omega_M$. Note that the Alfvén waves become linearly polarized only when the Coriolis force vanishes: when it is present, whatever its magnitude is, the modified Alfvén waves are circularly polarized.

II.c Liquid Metal and Hartmann Layer

(1) The presence of insulating plates imposes the boundary condition

$$\nabla \times \mathbf{b} = 0,$$

and thus $\mathbf{b} = \nabla \phi$. Since the problem depends on the variable y, we get $b_x(\pm \ell) = 0$.

(2) In the stationary regime the incompressible MHD equations reduce to

$$\mathbf{u} \cdot \nabla \mathbf{u} = -\nabla P_* + \mathbf{b} \cdot \nabla \mathbf{b} + \nu \, \Delta \mathbf{u},$$

$$\mathbf{u} \cdot \nabla \mathbf{b} = \mathbf{b} \cdot \nabla \mathbf{u} + \eta \, \Delta \mathbf{b}.$$

Since the problem depends only on the variable y and we have that the liquid metal flows in the x direction, one has $\mathbf{u} = (u_x, 0, 0)$ and $\mathbf{b} = (b_x, b_0, 0)$ with

$$\frac{d^2 u_x(y)}{dy^2} + \frac{b_0}{\nu} \frac{db_x(y)}{dy} = C,$$

$$\frac{d^2 b_x(y)}{dy^2} + \frac{b_0}{\eta} \frac{du_x(y)}{dy} = 0,$$

where the constant C is related to the pressure gradient by the relation $C\nu \equiv \partial P_* / \partial x$.

(3) By inserting the second expression into the first derivative with respect to y, one gets:

$$\frac{d^3 u_x(y)}{dy^3} - \frac{b_0^2}{\nu \eta} \frac{du_x(y)}{dy} = 0.$$

The solution is obtained by introducing a function of the type $u_x = \lambda \exp(\alpha y) + \mu \exp(-\beta y)$. With the boundary condition $u_x(\pm \ell) = 0$, one eventually gets

$$u_x(y) = A \left(1 - \frac{\cosh(H_a y / \ell)}{\cosh(H_a)} \right),$$

with the Hartmann dimensionless number $H_a \equiv b_0 \ell / \sqrt{\nu \eta}$.

(4) In Figure A1.2 we plot the normalized profile ($A = 1$, $\ell = 1$) of the solution $u_x(y)$ depending on the Hartmann number. We see that for a large H_a the fluid has an almost uniform velocity, while for a small H_a we tend to find the Poiseuille parabolic profile with the form $u_x(y) \propto (1 - (y/\ell)^2)$. This result shows that for a given liquid metal, it is the intensity of the applied magnetic field that will determine the profile.

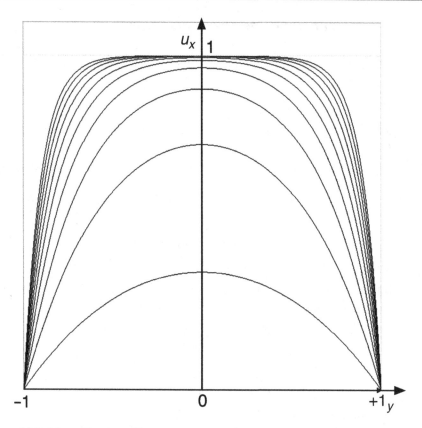

Figure A1.2 Normalized profile ($A = 1$, $\ell = 1$) of the solution $u_x(y)$ depending on the Hartmann number H_a going from 1 to 10 (from bottom to top).

II.d Magnetic Field Lines and the X-Point

To find the equation of a magnetic field line we use the relationship

$$\frac{dx}{B_x} = \frac{dy}{B_y}. \tag{1.1}$$

This expression comes from the condition $d\mathbf{r} \times \mathbf{B} = 0$ which is valid on a magnetic field line. The general solution is $y^2 = a^2 x^2 + c$, where c is a constant. An illustration of the solutions is given in Figure A1.3 for $a = 1$. The orientation of the field lines is easily obtained by looking at the direction of $\mathbf{B}$ at $x = 0$ or $y = 0$. The magnetic pressure force is $\mathbf{F_P} = (-1/\mu_0)(a^4 x, y)$ and the tension force is $\mathbf{F_T} = (a^2/\mu_0)(x, y)$. The current density is $\mathbf{j} = (1/\mu_0)(a^2 - 1)\mathbf{e_z}$, with $\mathbf{e_z}$, a unit vector. Therefore, the current is always null when $a = 1$ (the symmetric case).

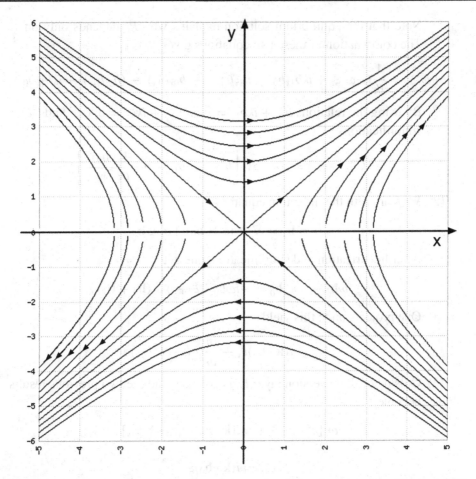

Figure A1.3 Magnetic field lines of $\mathbf{B} = (y, x)$. Note the presence of an X-point at the origin.

III.a Tearing Resistive Instability

(1) We start from the small-displacement equation

$$-\omega^2 \boldsymbol{\xi} = \frac{1}{\rho_0} \nabla(\boldsymbol{\xi} \cdot \nabla P_0) + (\nabla \times \mathbf{b}_1) \times \mathbf{b}_0 + (\nabla \times \mathbf{b}_0) \times \mathbf{b}_1,$$

where $\mathbf{b}_1 \equiv \nabla \times (\boldsymbol{\xi} \times \mathbf{b}_0)$ and $\mathbf{b}_0 = b_{0x}(y)\mathbf{e_x} + b_{0z}(y)\mathbf{e_z}$. After projection along y and z we find

$$-\omega^2 \xi_y = \frac{1}{\rho_0}(\xi_y P_0')' - b_{0x}b_{1x}' - b_{0z}b_{1z}' + ikb_{1y}b_{0z} - b_{1x}b_{0x}' - b_{1z}b_{0z}',$$

$$-\omega^2 \xi_z = \frac{ik}{\rho_0}\xi_y P_0' - ikb_{1x}b_{0x} + b_{1y}b_{0z}'.$$

Note that the equilibrium solution explains why P_0 depends only on y. The combination of these two equations gives

$$-\omega^2 \xi_y = \frac{1}{ik}(-\omega^2 \xi_z + ikb_{1x}b_{0x} - b_{1y}b_{0z}')' - (b_{0x}b_{1x})' - (b_{0z}b_{1z})' + ikb_{1y}b_{0z} .$$

By using the relations $b_{1y}' + ikb_{1z} = 0$ and $\xi_y' + ik\xi_z = 0$, one finds after simplifications

$$\omega^2[k^2\xi_y - \xi_y''] = i(\mathbf{k} \cdot \mathbf{b}_0)\left[b_{1y}'' - k^2 b_{1y} - b_{1y}\frac{(\mathbf{k} \cdot \mathbf{b}_0)''}{(\mathbf{k} \cdot \mathbf{b}_0)} \right] .$$

(2) We start with the induction equation

$$\dot{\mathbf{b}} = \dot{\mathbf{b}}_1 = \nabla \times (\dot{\boldsymbol{\xi}} \times \mathbf{b}_0) + \eta \, \Delta\mathbf{b} .$$

After linearization and simplification, one gets

$$-i\omega\mathbf{b}_1 - i\omega\xi_y\mathbf{b}_0' = -i\omega b_{0z}ik\boldsymbol{\xi} + \eta \, \Delta(\mathbf{b}_0 + \mathbf{b}_1) .$$

One projects on y; this yields

$$b_{1y} = i(\mathbf{k} \cdot \mathbf{b}_0)\xi_y + \frac{i\eta}{\omega}(b_{1y}'' - k^2 b_{1y}) .$$

(3) Outside of the inversion layer b_0 is constant and the two previous results become

$$\omega^2[k^2\xi_y - \xi_y''] = i(\mathbf{k} \cdot \mathbf{b}_0)[b_{1y}'' - k^2 b_{1y}] ,$$

$$b_{1y} = i(\mathbf{k} \cdot \mathbf{b}_0)\xi_y .$$

This gives us

$$\omega^2[k^2 b_{1y} - b_{1y}''] = -(\mathbf{k} \cdot \mathbf{b}_0)^2[b_{1y}'' - k^2 b_{1y}] ,$$

and eventually $\omega^2 = (\mathbf{k} \cdot \mathbf{b}_0)^2$. The stable solution is thus an Alfvén wave.

(4) With the proposed approximations, one has

$$\omega^2 \xi_y'' \simeq -i(\mathbf{k} \cdot \mathbf{b}_0)b_{1y}'' ,$$

$$(\mathbf{k} \cdot \mathbf{b}_0)\xi_y \simeq -\frac{\eta}{\omega}b_{1y}'' .$$

Hence we obtain the relation

$$\omega^2\frac{\xi_y}{\epsilon^2 L^2} \simeq i(\mathbf{k} \cdot \mathbf{b}_0)^2\frac{\omega}{\eta}\xi_y ,$$

and finally

$$\omega \simeq i\frac{(\mathbf{k} \cdot \mathbf{b}_0)^2(\epsilon L)^2}{\eta} = iA , \quad A \geq 0 .$$

This relationship is valid in the resistive layer. The solution therefore corresponds to a growing instability.

(5) We start from the expressions

$$(\mathbf{k} \cdot \mathbf{b}_0)\xi_y \simeq -\frac{\eta}{\omega}b_{1y}'',$$

and

$$\omega^2 \xi_y'' \simeq -i(\mathbf{k} \cdot \mathbf{b}_0)b_{1y}''.$$

This gives

$$(\mathbf{k} \cdot \mathbf{b}_0)\xi_y \simeq \frac{\eta}{\omega}\frac{\omega^2}{i(\mathbf{k} \cdot \mathbf{b}_0)}\xi_y'' \simeq \frac{-i\eta\omega}{(\mathbf{k} \cdot \mathbf{b}_0)}\frac{\xi_y}{(\epsilon L)^2},$$

and thus

$$(\mathbf{k} \cdot \mathbf{b}_0)^2(\epsilon L)^2 \simeq -i\eta\omega = \eta A.$$

With the proposed approximations, one obtains

$$\epsilon L \simeq \left(\frac{\eta A}{k^2(b_0'(0))^2}\right)^{1/4}.$$

(6) One starts from the following relations:

$$\omega^2 \xi_y'' = -i(\mathbf{k} \cdot \mathbf{b}_0)b_{1y}'',$$
$$b_{1y} = i(\mathbf{k} \cdot \mathbf{b}_0)\xi_y.$$

Hence

$$\frac{\omega^2\xi_y}{(\epsilon L)^2} \simeq -i(\mathbf{k} \cdot \mathbf{b}_0)b_{1y}'' \simeq -ikb_0'(0)Kb_{1y}(0) \simeq k^2(b_0'(0))^2(\epsilon L)K\xi_y,$$

and then

$$\omega^2 \simeq Kk^2(b_0'(0))^2\left(\frac{\eta A}{k^2(b_0'(0))^2}\right)^{3/4}.$$

Finally, a last operation allows us to express the solution in the form

$$A\tau_A \simeq (LK)^{4/5}(Lk)^{2/5}(\tau_A/\tau_R)^{3/5}.$$

The characteristic time of the tearing instability development therefore appears as intermediary compared with the two reference times. Note that this result can also be expressed with the Lundquist number $S = \tau_R/\tau_A$. This instability is in fact intimately linked to the magnetic reconnection process.

IV.a Model of Turbulent Diffusion

(1) The equation does not involve a linear dissipative term or an external force. It is therefore a model of energy diffusion in the inertial range.

(2) Using the spherical coordinates, we obtain

$$\frac{\partial e(\mathbf{k})}{\partial t} = -\frac{1}{k^2}\frac{\partial(k^2 F_k)}{\partial k}.$$

(3) We see that the dimension of the diffusion coefficient is

$$[D] = \left[\frac{F_r k}{e}\right] = \left[\frac{ke\,k}{\tau\,e}\right] = \left[\frac{k^2}{\tau}\right].$$

D is therefore measured in $m^{-2}\,s^{-1}$.

(4) One has

$$\tau = \frac{\ell}{v_\ell} = \frac{\ell}{\sqrt{ek^3}} = k^{-5/2}e^{-1/2}.$$

(5) With the proposed notations, this yields

$$\frac{\partial e(\mathbf{k})}{\partial t} = \frac{\partial}{\partial t}\left(\frac{E(k)}{4\pi k^2}\right) = \frac{1}{k^2}\frac{\partial}{\partial k}\left(k^2 D\frac{\partial}{\partial k}\left(\frac{E}{4\pi k^2}\right)\right).$$

On the other hand, one has

$$D = \frac{k^2}{\tau} = k^{9/2}\sqrt{\frac{E}{4\pi k^2}}.$$

Hence finally, after rearrangement,

$$\frac{\partial E(k)}{\partial t} = C\frac{\partial}{\partial k}\left(k^{11/2}\sqrt{E(k)}\frac{\partial(E(k)/k^2)}{\partial k}\right).$$

(6) We seek the power-law solution such that

$$\varepsilon \sim k^{11/2}\sqrt{E(k)}\frac{\partial(E(k)/k^2)}{\partial k} = \text{constant}.$$

This gives the Kolmogorov spectrum $E(k) \sim k^{-5/3}$.

(7) In the MHD case we have to introduce a new characteristic time $\tau = \ell B_0/v_\ell^2$. Hence the form of the isotropic diffusion equation in MHD becomes

$$\frac{\partial E(k)}{\partial t} = \tilde{C}\frac{\partial}{\partial k}\left(k^6 E(k)\frac{\partial(E(k)/k^2)}{\partial k}\right),$$

whose power-law constant-flux solution is the IK spectrum $E(k) \sim k^{-3/2}$.

IV.b Whistler Turbulence

(1) We recall that the electron MHD equation governing whistler turbulence can be written as:

$$\frac{\partial \mathbf{b}}{\partial t} = -d_i \, \nabla \times ((\nabla \times \mathbf{b}) \times \mathbf{b}) + \eta \, \Delta \mathbf{b},$$

where **b** is the magnetic field normalized with respect to a velocity and d_i is the ion inertial length. The transfer time of energy is therefore $\tau_{nl} = \ell^2/(d_i b_\ell)$.

(2) In the inertial range we have the relation

$$\varepsilon \sim \frac{E(k)k}{\tau_{tr}} \sim d_i k^{7/2} E^{3/2}(k),$$

hence the isotropic magnetic energy spectrum is given by $E(k) = (\varepsilon/d_i)^{2/3} k^{-7/3}$.

(3) In presence of a uniform magnetic field $\mathbf{b}_0 = b_0 \mathbf{e}_z$, one obtains

$$\frac{\partial \mathbf{b}}{\partial t} + d_i b_0 \frac{\partial}{\partial z}(\nabla \times \mathbf{b}) = -d_i \, \nabla \times ((\nabla \times \mathbf{b}) \times \mathbf{b}) + \eta \, \Delta \mathbf{b},$$

which gives the whistler time $\tau_w = 1/(k_\| k_\perp d_i b_0)$. Considering stochastic collisions of whistler wave-packets, one gets the transfer time

$$\tau_{tr} = \frac{\tau_{nl}^2}{\tau_w} = \frac{k_\| b_0}{k_\perp^3 d_i b_\ell^2}.$$

(4) From the transfer time, we can deduce the axisymmetric anisotropic whistler turbulence spectrum. In the inertial zone, one has

$$\varepsilon \sim \frac{(E(k_\perp, k_\|) k_\perp k_\|)^2}{\tau_{tr}} \sim \frac{E^2(k_\perp, k_\|) k_\perp^5 k_\| d_i}{b_0},$$

hence, eventually,

$$E(k_\perp, k_\|) \sim \sqrt{\frac{\varepsilon b_0}{d_i}} k_\perp^{-5/2} k_\|^{-1/2}.$$

Appendix 2

Formulary

Vector Identities

- Multiple products:

$$(\mathbf{A} \times \mathbf{B}) \cdot (\mathbf{C} \times \mathbf{D}) = (\mathbf{A} \cdot \mathbf{C})(\mathbf{B} \cdot \mathbf{D}) - (\mathbf{A} \cdot \mathbf{D})(\mathbf{B} \cdot \mathbf{C})$$
$$\mathbf{A} \cdot (\mathbf{B} \times \mathbf{C}) = \mathbf{B} \cdot (\mathbf{C} \times \mathbf{A}) = \mathbf{C} \cdot (\mathbf{A} \times \mathbf{B})$$
$$\mathbf{A} \times (\mathbf{B} \times \mathbf{C}) = \mathbf{B}(\mathbf{A} \cdot \mathbf{C}) - \mathbf{C}(\mathbf{A} \cdot \mathbf{B})$$

- Rules for products with derivatives:

$$\nabla(fg) = f(\nabla g) + g(\nabla f)$$
$$\nabla(\mathbf{A} \cdot \mathbf{B}) = \mathbf{A} \times (\nabla \times \mathbf{B}) + \mathbf{B} \times (\nabla \times \mathbf{A}) + (\mathbf{A} \cdot \nabla)\mathbf{B} + (\mathbf{B} \cdot \nabla)\mathbf{A}$$
$$\nabla \cdot (f\mathbf{A}) = f(\nabla \cdot \mathbf{A}) + \mathbf{A} \cdot (\nabla f)$$
$$\nabla \cdot (\mathbf{A} \times \mathbf{B}) = \mathbf{B} \cdot (\nabla \times \mathbf{A}) - \mathbf{A} \cdot (\nabla \times \mathbf{B})$$
$$\nabla \times (f\mathbf{A}) = f(\nabla \times \mathbf{A}) - \mathbf{A} \times (\nabla f)$$
$$\nabla \times (\mathbf{A} \times \mathbf{B}) = (\mathbf{B} \cdot \nabla)\mathbf{A} - (\mathbf{A} \cdot \nabla)\mathbf{B} + \mathbf{A}(\nabla \cdot \mathbf{B}) - \mathbf{B}(\nabla \cdot \mathbf{A})$$

- Rules for second-order derivatives:

$$\nabla \cdot (\nabla \times \mathbf{A}) = 0$$
$$\nabla \times (\nabla f) = \mathbf{0}$$
$$\nabla \times (\nabla \times \mathbf{A}) = \nabla(\nabla \cdot \mathbf{A}) - \Delta\mathbf{A}$$

Vectorial derivatives

We list here the vectorial derivatives in the three usual coordinate systems (see Figure A2.1).

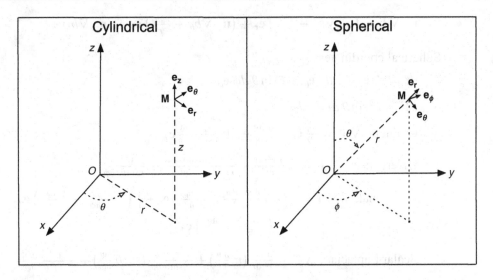

Figure A2.1 Systems of cylindrical polar and spherical coordinates.

- Cartesian coordinates:

$$d\boldsymbol{\ell} = dx\,\mathbf{e_x} + dy\,\mathbf{e_y} + dz\,\mathbf{e_z}$$

$$dV = dx\,dy\,dz$$

Gradient: $\nabla\Psi = \frac{\partial\Psi}{\partial x}\mathbf{e_x} + \frac{\partial\Psi}{\partial y}\mathbf{e_y} + \frac{\partial\Psi}{\partial z}\mathbf{e_z}$

Divergence: $\nabla\cdot\mathbf{u} = \frac{\partial u_x}{\partial x} + \frac{\partial u_y}{\partial y} + \frac{\partial u_z}{\partial z}$

Rotational: $\nabla\times\mathbf{u} = \left(\frac{\partial u_z}{\partial y} - \frac{\partial u_y}{\partial z}\right)\mathbf{e_x} + \left(\frac{\partial u_x}{\partial z} - \frac{\partial u_z}{\partial x}\right)\mathbf{e_y} + \left(\frac{\partial u_y}{\partial x} - \frac{\partial u_x}{\partial y}\right)\mathbf{e_z}$

Scalar Laplacian: $\Delta\Psi = \frac{\partial^2\Psi}{\partial x^2} + \frac{\partial^2\Psi}{\partial y^2} + \frac{\partial^2\Psi}{\partial z^2}$

Vectorial case: $\Delta\mathbf{u} = \Delta u_x\,\mathbf{e_x} + \Delta u_y\,\mathbf{e_y} + \Delta u_z\,\mathbf{e_z}$

- Cylindrical coordinates:

$$d\boldsymbol{\ell} = dr\,\mathbf{e_r} + r\,d\theta\,\mathbf{e_\theta} + dz\,\mathbf{e_z}$$

$$dV = r\,dr\,d\theta\,dz$$

Gradient: $\nabla\Psi = \frac{\partial\Psi}{\partial r}\mathbf{e_r} + \frac{1}{r}\frac{\partial\Psi}{\partial\theta}\mathbf{e_\theta} + \frac{\partial\Psi}{\partial z}\mathbf{e_z}$

Divergence: $\nabla\cdot\mathbf{u} = \frac{1}{r}\frac{\partial}{\partial r}(r u_r) + \frac{1}{r}\frac{\partial u_\theta}{\partial\theta} + \frac{\partial u_z}{\partial z}$

Rotational: $\nabla\times\mathbf{u} = \left(\frac{1}{r}\frac{\partial u_z}{\partial\theta} - \frac{\partial u_\theta}{\partial z}\right)\mathbf{e_r} + \left(\frac{\partial u_r}{\partial z} - \frac{\partial u_z}{\partial r}\right)\mathbf{e_\theta} + \frac{1}{r}\left(\frac{\partial}{\partial r}(r u_\theta) - \frac{\partial u_r}{\partial\theta}\right)\mathbf{e_z}$

Scalar Laplacian: $\Delta\Psi = \frac{1}{r}\frac{\partial}{\partial r}\left(r\frac{\partial\Psi}{\partial r}\right) + \frac{1}{r^2}\frac{\partial^2\Psi}{\partial\theta^2} + \frac{\partial^2\Psi}{\partial z^2}$

Vectorial case: $\Delta\mathbf{u} = \left(\Delta u_r - \frac{u_r}{r^2} - \frac{2}{r^2}\frac{\partial u_\theta}{\partial\theta}\right)\mathbf{e_r} + \left(\Delta u_\theta - \frac{u_\theta}{r^2} + \frac{2}{r^2}\frac{\partial u_r}{\partial\theta}\right)\mathbf{e_\theta}$
$\qquad\qquad + \Delta u_z\,\mathbf{e_z}$

$$\mathbf{u} \cdot \nabla \mathbf{b} = \left(\mathbf{u} \cdot \nabla b_r - \tfrac{u_\theta b_\theta}{r}\right) \mathbf{e_r} + \left(\mathbf{u} \cdot \nabla b_\theta + \tfrac{u_\theta b_r}{r}\right) \mathbf{e_\theta} + (\mathbf{u} \cdot \nabla b_z)\mathbf{e_z}$$

- Spherical coordinates:

$$d\boldsymbol{\ell} = dr\, \mathbf{e_r} + r\, d\theta\, \mathbf{e_\theta} + r \sin\theta\, d\phi\, \mathbf{e_\phi}$$

$$d\mathcal{V} = r^2 \sin\theta\, dr\, d\theta\, d\phi$$

Gradient: $\nabla\Psi = \dfrac{\partial\Psi}{\partial r}\mathbf{e_r} + \dfrac{1}{r}\dfrac{\partial\Psi}{\partial\theta}\mathbf{e_\theta} + \dfrac{1}{r\sin\theta}\dfrac{\partial\Psi}{\partial\phi}\mathbf{e_\phi}$

Divergence: $\nabla \cdot \mathbf{u} = \dfrac{1}{r^2}\dfrac{\partial(r^2 u_r)}{\partial r} + \dfrac{1}{r\sin\theta}\dfrac{\partial(\sin\theta\, u_\theta)}{\partial\theta} + \dfrac{1}{r\sin\theta}\dfrac{\partial u_\phi}{\partial\phi}$

Rotational: $\nabla \times \mathbf{u} = \dfrac{1}{r\sin\theta}\left(\dfrac{\partial(\sin\theta\, u_\phi)}{\partial\theta} - \dfrac{\partial u_\theta}{\partial\phi}\right)\mathbf{e_r} + \dfrac{1}{r}\left(\dfrac{1}{\sin\theta}\dfrac{\partial u_r}{\partial\phi} - \dfrac{\partial(r u_\phi)}{\partial r}\right)\mathbf{e_\theta}$
$$+ \dfrac{1}{r}\left(\dfrac{\partial(r u_\theta)}{\partial r} - \dfrac{\partial u_r}{\partial\theta}\right)\mathbf{e_\phi}$$

Scalar Laplacian: $\Delta\Psi = \dfrac{1}{r^2}\dfrac{\partial}{\partial r}\left(r^2\dfrac{\partial\Psi}{\partial r}\right) + \dfrac{1}{r^2\sin\theta}\dfrac{\partial}{\partial\theta}\left(\sin\theta\dfrac{\partial\Psi}{\partial\theta}\right) + \dfrac{1}{r^2\sin^2\theta}\dfrac{\partial^2\Psi}{\partial\phi^2}$

Vectorial case: $\Delta\mathbf{u} = \left(\Delta u_r - \dfrac{2}{r^2}\left(\dfrac{1}{\sin\theta}\dfrac{\partial(\sin\theta\, u_\theta)}{\partial\theta} + \dfrac{1}{\sin\theta}\dfrac{\partial u_\phi}{\partial\phi} + u_r\right)\right)\mathbf{e_r}$
$$+ \left(\Delta u_\theta + \dfrac{1}{r^2}\left(\dfrac{\partial 2u_r}{\partial\theta} - \dfrac{u_\theta}{\sin^2\theta} - \dfrac{2\cos\theta}{\sin^2\theta}\dfrac{\partial u_\phi}{\partial\phi}\right)\right)\mathbf{e_\theta}$$
$$+ \left(\Delta u_\phi + \dfrac{1}{r^2}\left(\dfrac{2}{\sin\theta}\dfrac{\partial u_r}{\partial\phi} + \dfrac{2\cos\theta}{\sin^2\theta}\dfrac{\partial u_\theta}{\partial\phi} - \dfrac{u_\phi}{\sin^2\theta}\right)\right)\mathbf{e_\phi}$$

$$\mathbf{u} \cdot \nabla \mathbf{b} = \left(\mathbf{u} \cdot \nabla b_r - \tfrac{u_\theta b_\theta + u_\phi b_\phi}{r}\right) \mathbf{e_r} + \left(\mathbf{u} \cdot \nabla b_\theta + \tfrac{u_\theta b_r - u_\phi b_\phi \cot\theta}{r}\right) \mathbf{e_\theta}$$
$$+ \left(\mathbf{u} \cdot \nabla b_\phi + \tfrac{u_\phi b_r + u_\phi b_\theta \cot\theta}{r}\right) \mathbf{e_\phi}$$

Fundamental theorems

- Gradient theorem:

$$\int_a^b (\nabla f) \cdot d\boldsymbol{\ell} = f(b) - f(a)$$

- Divergence (or Ostrogradsky) theorem:

$$\iiint (\nabla \cdot \mathbf{A}) d\mathcal{V} = \oiint \mathbf{A} \cdot d\mathcal{S}$$

- Rotational (or Stokes) theorem:

$$\iint (\nabla \times \mathbf{A}) \cdot d\mathcal{S} = \oint \mathbf{A} \cdot d\boldsymbol{\ell}$$

References

Abramenko, V. I. 2005. Relationship between magnetic power spectrum and flare productivity in solar active regions. *Astrophys. J.*, **629**, 1141–1149.

Abramenko, V. I., Yurchyshyn, V. B., Wang, H., Spirock, T. J., and Goode, P. R. 2002. Scaling behavior of structure functions of the longitudinal magnetic field in active regions on the Sun. *Astrophys. J.*, **577**, 487–495.

Alexandrova, O., Lacombe, C., Mangeney, A., Grappin, R., and Maksimovic, M. 2012. Solar wind turbulent spectrum at plasma kinetic scales. *Astrophys. J.*, **760**, 121.

Alfvén, H. 1942. Existence of electromagnetic–hydrodynamic waves. *Nature*, **150**, 405–406.

Alfvén, H. 1942. On the existence of electromagnetic–hydrodynamic waves. *Ark. Mat. Astron. och Fys.*, **29**, 1–7.

Allen, T. K., Baker, W. R., Pyle, R. V., and Wilcox, J. M. 1959. Experimental generation of plasma Alfvén waves. *Phys. Rev. Lett.*, **2**, 383–384.

Antonia, R. A., and Burattini, P. 2006. Approach to the 4/5 law in homogeneous isotropic turbulence. *J. Fluid Mech.*, **550**, 175–184.

Antonia, R. A., Ould-Rouis, M., Anselmet, F., and Zhu, Y. 1997. Analogy between predictions of Kolmogorov and Yaglom. *J. Fluid Mech.*, **332**, 395–409.

Arzoumanian, D., André, P., Didelon, P. *et al.* 2011. Characterizing interstellar filaments with Herschel in IC 5146. *Astron. Astrophys.*, **529**, L6.

Aubourg, Q., and Mordant, N. 2015. Nonlocal resonances in weak turbulence of gravity–capillary waves. *Phys. Rev. Lett.*, **114**, 144501.

Aulanier, G., DeLuca, E. E., Antiochos, S. K., McMullen, R. A., and Golub, L. 2000. The topology and evolution of the Bastille Day flare. *Astrophys. J.*, **540**, 1126–1142.

Aulanier, G., Pariat, E., Démoulin, P., and DeVore, C. R. 2006. Slip-running reconnection in quasi-separatrix layers. *Sol. Phys.*, **238**, 347–376.

Aunai, N., Belmont, G., and Smets, R. 2011. Ion acceleration in antiparallel collisionless magnetic reconnection: Kinetic and fluid aspects. *Comptes Rendus Phys.*, **12**, 141–150.

Balbus, S. A., and Hawley, J. F. 1991. A powerful local shear instability in weakly magnetized disks. I – Linear analysis. *Astrophys. J.*, **376**, 214–233.

Banerjee, S., and Galtier, S. 2013. Exact relation with two-point correlation functions and phenomenological approach for compressible magnetohydrodynamic turbulence. *Phys. Rev. E*, **87**(1), 013019.

Baumjohann, W., and Treumann, R. A. 1996. *Basic Space Plasma Physics*. Imperial College Press.

Belmont, G., Grappin, R., Mottez, F., Pantellini, F., and Pelletier, G. 2013. *Collisionless Plasmas in Astrophysics*. Wiley.

Benney, J., and Newell, A. C. 1966. Random wave closures. *Studies Appl. Math.*, **48**, 29–53.

Beresnyak, A. 2011. Spectral slope and Kolmogorov constant of MHD turbulence. *Phys. Rev. Lett.*, **106**(7), 075001.

Berger, M. A., and Field, G. B. 1984. The topological properties of magnetic helicity. *J. Fluid Mech.*, **147**, 133–148.

Berhanu, M., Monchaux, R., Fauve, S. *et al.* 2007. Magnetic field reversals in an experimental turbulent dynamo. *Europhys. Lett.*, **77**, 59001.

Bewley, G. P., Paoletti, M. S., Sreenivasan, K. R., and Lathrop, D. P. 2008. Characterization of reconnecting vortices in superfluid helium. *Proc. Nat. Acad. Sci.*, **105**, 13707–13710.

Bhattacharjee, A. 2004. Impulsive magnetic reconnection in the Earth's magnetotail and the solar corona. *Ann. Rev. Astron. Astrophys.*, **42**, 365–384.

Bigot, B., Galtier, S., and Politano, H. 2008a. An anisotropic turbulent model for solar coronal heating. *Astron. Astrophys.*, **490**, 325–337.

Bigot, B., Galtier, S., and Politano, H. 2008b. Development of anisotropy in incompressible magnetohydrodynamic turbulence. *Phys. Rev. E*, **78**(6), 066301.

Biskamp, D. 1986. Magnetic reconnection via current sheets. *Phys. Fluids*, **29**, 1520–1531.

Biskamp, D. 2000. *Magnetic Reconnection in Plasmas*. Cambridge University Press.

Biskamp, D. 2003. *Magnetohydrodynamic Turbulence*. Cambridge University Press.

Biskamp, D., Schwarz, E., and Drake, J. F. 1996. Two-dimensional electron magnetohydrodynamic turbulence. *Phys. Rev. Lett.*, **76**, 1264–1267.

Boldyrev, S. 2006. Spectrum of magnetohydrodynamic turbulence. *Phys. Rev. Lett.*, **96**(11), 115002.

Boltzmann, L. 1872. Weitere Studien über das Wärmegleichgewicht unter Gasmolekülen. *Wiener Berichte*, **66**, 275–370.

Bostick, W. H., and Levine, M. A. 1952. Experimental demonstration in the laboratory of the existence of magneto-hydrodynamic waves in ionized helium. *Phys. Rev.*, **87**, 671–671.

Braginskii, S. I. 1965. Transport processes in a plasma. *Rev. Plasma Phys.*, **1**, 205–311.

Brandenburg, A., and Subramanian, K. 2005. Astrophysical magnetic fields and nonlinear dynamo theory. *Phys. Rep.*, **417**, 1–209.

Buchlin, E., and Velli, M. 2007. Shell models of reduced MHD turbulence and the heating of solar coronal loops. *Astrophys. J.*, **662**, 701–714.

Bullard, Sir, E. 1955. The stability of a homopolar dynamo. *Proc. Cambridge Phil. Soc.*, **51**, 744–760.

Busse, F. H., and Wicht, J. 1992. A simple dynamo caused by conductivity variations. *Geophys. Astrophys. Fluid Dyn.*, **64**, 135–144.

Campagne, A., Gallet, B., Moisy, F., and Cortet, P.-P. 2015. Disentangling inertial waves from eddy turbulence in a forced rotating-turbulence experiment. *Phys. Rev. E*, **91**(4), 043016.

Canou, A., and Amari, T. 2010. A twisted flux rope as the magnetic structure of a filament in Active Region 10953 observed by Hinode. *Astrophys. J.*, **715**, 1566–1574.

Carbone, V., Bruno, R., Sorriso-Valvo, L., and Lepreti, F. 2004. Intermittency of magnetic turbulence in slow solar wind. *Plan. Space Sci.*, **52**, 953–956.

Chabrier, G., and Hennebelle, P. 2011. Dimensional argument for the impact of turbulent support on the stellar initial mass function. *Astron. Astrophys.*, **534**, A106.

Chandran, B. D. G. 2010. Alfvén-wave turbulence and perpendicular ion temperatures in coronal holes. *Astrophys. J.*, **720**, 548–554.

Chandrasekhar, S. 1960. The stability of non-dissipative Couette flow in hydromagnetics. *Proc. Nat. Acad. Sci.*, **46**, 253–257.

Chapman, S. 1931. The absorption and dissociative or ionizing effect of monochromatic radiation in an atmosphere on a rotating Earth. *Proc. Phys. Soc.*, **43**, 26–45.

Cho, J., and Vishniac, E. T. 2000. The anisotropy of magnetohydrodynamic Alfvénic turbulence. *Astrophys. J.*, **539**, 273–282.

Clark di Leoni, P., Cobelli, P. J., and Mininni, P. D. 2014. Wave turbulence in shallow water models. *Phys. Rev. E*, **89**(6), 063025.

Cook, A. E., and Roberts, P. H. 1970. The Rikitake two-disc dynamo system. *Proc. Camb. Phil. Soc.*, **68**, 547.

Cooper, C. M., Wallace, J., Brookhart, M. *et al.* 2014. The Madison plasma dynamo experiment: A facility for studying laboratory plasma astrophysics. *Phys. Plasmas*, **21**(1), 013505.

Cowling, T. G. 1933. The magnetic field of sunspots. *Month. Not. Roy. Astron. Soc.*, **94**, 39–48.

Crémer, P., and Alemany, A. 1981. Aspects expérimentaux du brassage électromagnétique en creuset. *J. Méc. Appl.*, **5**, 37–50.

Daughton, W., Roytershteyn, V., Albright, B. J. *et al.* 2009. Transition from collisional to kinetic regimes in large-scale reconnection layers. *Phys. Rev. Lett.*, **103**(6), 065004.

Davidson, P. A. 2001. *An Introduction to Magnetohydrodynamics*. Cambridge University Press.

Davidson, P. A. 2004. *Turbulence: An Introduction for Scientists and Engineers*. Cambridge University Press.

De Pontieu, B., McIntosh, S. W., Carlsson, M. *et al.* 2007. Chromospheric Alfvénic waves strong enough to power the solar wind. *Science*, **318**, 1574–1577.

Diamond, P. H., Itoh, S.-I., and Itoh, K. 2010. *Modern Plasma Physics*. Cambridge University Press.

Dubrulle, B. 1994. Intermittency in fully developed turbulence: Log-Poisson statistics and generalized scale covariance. *Phys. Rev. Lett.*, **73**, 959–962.

Elsässer, W. M. 1950. The hydromagnetic equations. *Phys. Rev.*, **79**, 183–183.

Falcon, É., Laroche, C., and Fauve, S. 2007. Observation of gravity-capillary wave turbulence. *Phys. Rev. Lett.*, **98**(9), 094503.

Falthammar, C. 2007. The discovery of magnetohydrodynamic waves. *J. Atmos. Solar–Terrestrial Phys.*, **69**, 1604–1608.

Ferreira, J. 1997. Magnetically-driven jets from Keplerian accretion discs. *Astron. Astrophys.*, **319**, 340–359.

Feynman, R. P., Leighton, R. B., and Sands, M. 1964. *The Feynman Lectures on Physics. Mainly Electromagnetism and Matter. Volume 2*. Addison-Wesley Publishing Company.

Frisch, U. 1995. *Turbulence. The Legacy of A. N. Kolmogorov*. Cambridge University Press.

Frisch, U., Pouquet, A., Leorat, J., and Mazure, A. 1975. Possibility of an inverse cascade of magnetic helicity in magnetohydrodynamic turbulence. *J. Fluid Mech.*, **68**, 769–778.

Frisch, U., Sulem, P.-L., and Nelkin, M. 1978. A simple dynamical model of intermittent fully developed turbulence. *J. Fluid Mech.*, **87**, 719–736.

Gailitis, A., Lielausis, O., Dement'ev, S. *et al.* 2000. Detection of a flow induced magnetic field eigenmode in the Riga Dynamo Facility. *Phys. Rev. Lett.*, **84**, 4365–4368.

Galanti, B., and Tsinober, A. 2004. Is turbulence ergodic? *Phys. Lett. A*, **330**, 173–180.

Galtier, S. 2006. Wave turbulence in incompressible Hall magnetohydrodynamics. *J. Plasma Phys.*, **72**, 721–769.

Galtier, S. 2008. Von Kármán–Howarth equations for Hall magnetohydrodynamic flows. *Phys. Rev. E*, **77**(1), 015302.

Galtier, S. 2012. Kolmogorov vectorial law for solar wind turbulence. *Astrophys. J.*, **746**, 184.

Galtier, S. 2014. Weak turbulence theory for rotating magnetohydrodynamics and planetary flows. *J. Fluid Mech.*, **757**, 114–154.

Galtier, S., and Banerjee, S. 2011. Exact relation for correlation functions in compressible isothermal turbulence. *Phys. Rev. Lett.*, **107**(13), 134501.

Galtier, S., Politano, H., and Pouquet, A. 1997. Self-similar energy decay in magneto-hydrodynamic turbulence. *Phys. Rev. Lett.*, **79**, 2807–2810.

Galtier, S., Nazarenko, S. V., Newell, A. C., and Pouquet, A. 2000. A weak turbulence theory for incompressible magnetohydrodynamics. *J. Plasma Phys.*, **63**, 447–488.

Galtier, S., Pouquet, A., and Mangeney, A. 2005. On spectral scaling laws for incompressible anisotropic magnetohydrodynamic turbulence. *Phys. Plasmas*, **12**(9), 092310.

Glatzmaier, G. A., and Roberts, P. H. 1995. A three-dimensional self-consistent computer simulation of a geomagnetic field reversal. *Nature*, **377**(Sept.), 203–209.

Goedbloed, J. P., Keppens, R., and Poedts, S. 2010. *Advanced Magnetohydrodynamics*. Cambridge University Press.

Goldreich, P., and Julian, W. H. 1969. Pulsar electrodynamics. *Astrophys. J.*, **157**, 869–880.

Goldreich, P., and Sridhar, S. 1995. Toward a theory of interstellar turbulence. 2: Strong Alfvénic turbulence. *Astrophys. J.*, **438**, 763–775.

Grappin, R., Frisch, U., Pouquet, A., and Leorat, J. 1982. Alfvénic fluctuations as asymptotic states of MHD turbulence. *Astron. Astrophys.*, **105**, 6–14.

Grauer, R., Krug, J., and Marliani, C. 1994. Scaling of high-order structure functions in magnetohydrodynamic turbulence. *Phys. Lett. A*, **195**, 335–338.

Haines, M. G., Lepell, P. D., Coverdale, C. A. *et al.* 2006. Ion viscous heating in a magnetohydrodynamically unstable Z Pinch at over 2×10^9 kelvin. *Phys. Rev. Lett.*, **96**, 075003.

Heisenberg, W. 1948. Zur statistischen Theorie der Turbulenz. *Z. Phys.*, **124**, 628–657.

Heyer, M. H., and Brunt, C. M. 2004. The universality of turbulence in galactic molecular clouds. *Astrophys. J.*, **615**, L45–L48.

Heyvaerts, J., and Priest, E. R. 1983. Coronal heating by phase-mixed shear Alfvén waves. *Astron. Astrophys.*, **117**, 220–234.

Higdon, J. C. 1984. Density fluctuations in the interstellar medium: Evidence for anisotropic magnetogasdynamic turbulence. I – Model and astrophysical sites. *Astrophys. J.*, **285**, 109–123.

Horbury, T. S., and Balogh, A. 1997. Structure function measurements of the intermittent MHD turbulent cascade. *Nonlin. Proc. Geophys.*, **4**, 185–199.

Hunana, P., Laveder, D., Passot, T., Sulem, P. L., and Borgogno, D. 2011. Reduction of compressibility and parallel transfer by Landau damping in turbulent magnetized plasmas. *Astrophys. J.*, **743**, 128.

Iroshnikov, R. S. 1964. Turbulence of a conducting fluid in a strong magnetic field. *Soviet Astron.*, **7**, 566–571.

Ji, H., Yamada, M., Hsu, S. *et al.* 1999. Magnetic reconnection with Sweet–Parker characteristics in two-dimensional laboratory plasmas. *Phys. Plasmas*, **6**, 1743–1750.

Kiyani, K. H., Chapman, S. C., Khotyaintsev, Y. V., Dunlop, M. W., and Sahraoui, F. 2009. Global scale-invariant dissipation in collisionless plasma turbulence. *Phys. Rev. Lett.*, **103**, 075006.

Kiyani, K. H., Osman, K. T., and Chapman, S. C. 2015. Dissipation and heating in solar wind turbulence: from the macro to the micro and back again. *Phil. Trans. R. Soc. Lond. A*, **373**(2041), 1–10.

Kolmogorov, A. N. 1941. Dissipation of energy in locally isotropic turbulence. *Dokl. Akad. Nauk SSSR*, **32**, 16–18.

Kolmogorov, A. N. 1962. A refinement of previous hypotheses concerning the local structure of turbulence in a viscous incompressible fluid at high Reynolds number. *J. Fluid Mech.*, **13**, 82–85.

Kraichnan, R. H. 1958. Irreversible statistical mechanics of incompressible hydromagnetic turbulence. *Phys. Rev.*, **109**, 1407–1422.

Kraichnan, R. H. 1965. Inertial-range spectrum of hydromagnetic turbulence. *Phys. Fluids*, **8**, 1385–1387.

Kraichnan, R. H. 1967. Inertial ranges in two-dimensional turbulence. *Phys. Fluids*, **10**, 1417–1423.

Kritsuk, A. G., Norman, M. L., Padoan, P., and Wagner, R. 2007. The statistics of supersonic isothermal turbulence. *Astrophys. J.*, **665**, 416–431.

Kruskal, M., and Schwarzschild, M. 1954. Some instabilities of a completely ionized plasma. *R. Soc. Lond. Proc. A*, **223**, 348–360.

Landau, L. D. 1946. On the vibrations of the electronic plasma. *J. Phys. USSR*, **10**, 25–34.

Langmuir, I. 1928. Oscillations in ionized gases. *Proc. Nat. Acad. Sci.*, **14**, 627.

Lehnert, B. 1954. Magneto-hydrodynamic waves in liquid sodium. *Phys. Rev.*, **94**, 815–824.

Leith, C. E. 1969. Diffusion approximation to spectral transfer in homogeneous turbulence. *Phys. Fluids*, **12**, 285.

Lorenz, E. N. 1963. Deterministic nonperiodic flow. *J. Atmos. Sci.*, **20**, 130–141.

Lundquist, S. 1949. Experimental investigations of magneto-hydrodynamic waves. *Phys. Rev.*, **76**, 1805–1809.

Luo, Q. Y., and Wu, D. J. 2010. Observations of anisotropic scaling of solar wind turbulence. *Astrophys. J. Lett.*, **714**, L138–L141.

Mandt, M. E., Denton, R. E., and Drake, J. F. 1994. Transition to whistler mediated magnetic reconnection. *Geophys. Res. Lett.*, **21**, 73–76.

Maxwell, J. C. 1873. *A Treatise on Electricity and Magnetism.* Clarendon Press.

Meyrand, R., and Galtier, S. 2012. Spontaneous chiral symmetry breaking of Hall magnetohydrodynamic turbulence. *Phys. Rev. Lett.*, **109**, 194501.

Meyrand, R., and Galtier, S. 2013. Anomalous $k_\perp^{-8/3}$ spectrum in electron magnetohydrodynamic turbulence. *Phys. Rev. Lett.*, **111**(26), 264501.

Meyrand, R., Kiyani, K. H., and Galtier, S. 2015. Weak magnetohydrodynamic turbulence and intermittency. *J. Fluid Mech. Rapids*, **770**, R1.

Mininni, P. D., Gómez, D. O., and Mahajan, S. M. 2005. Direct simulations of helical Hall–MHD turbulence and dynamo action. *Astrophys. J.*, **619**, 1019–1027.

Moffatt, H. K. 1969. The degree of knottedness of tangled vortex lines. *J. Fluid Mech.*, **35**, 117–129.

Moffatt, H. K. 1970. Dynamo action associated with random inertial waves in a rotating conducting fluid. *J. Fluid Mech.*, **44**, 705–719.

Moffatt, H. K. 1978. *Magnetic Field Generation in Electrically Conducting Fluids.* Cambridge University Press.

Monchaux, R., Berhanu, M., Bourgoin, M. *et al.* 2007. Generation of a magnetic field by dynamo action in a turbulent flow of liquid sodium. *Phys. Rev. Lett.*, **98**, 044502.

Mouhot, C., and Villani, C. 2010. Landau damping. *J. Math. Phys.*, **51**, 015204.

Mozer, F. S., Bale, S. D., and Phan, T. D. 2002. Evidence of diffusion regions at a subsolar magnetopause crossing. *Phys. Rev. Lett.*, **89**(1), 015002.

Müller, W.-C. 2009. Magnetohydrodynamic turbulence. Pages 223–254 of Hillebrandt, W., and Kupka, F. (eds.), *Interdisciplinary Aspects of Turbulence.* Springer Verlag.

Münch, G. 1958. Internal motions in the Orion nebula. *Rev. Mod. Phys.*, **30**, 1035–1041.

Nataf, H.-C., and Gagnière, N. 2008. On the peculiar nature of turbulence in planetary dynamos. *Comptes Rendus Phys.*, **9**, 702–710.

Nazarenko, S. 2011. *Wave Turbulence.* Springer Verlag.

Noether, E. 1918. Invariante variationsprobleme. *Nachr. König. Gesellsch. Wiss. Göttingen, Math. Phys. Klasse*, **1**, 235–237.

Nore, C., Brachet, M. E., Politano, H., and Pouquet, A. 1997. Dynamo action in the Taylor–Green vortex near threshold. *Phys. Plasmas*, **4**, 1–3.

Orszag, S. A. 1970. Analytical theories of turbulence. *J. Fluid Mech.*, **41**, 363–386.

Oughton, S., Priest, E. R., and Matthaeus, W. H. 1994. The influence of a mean magnetic field on three-dimensional magnetohydrodynamic turbulence. *J. Fluid Mech.*, **280**, 95–117.

Paret, J., and Tabeling, P. 1998. Intermittency in the two-dimensional inverse cascade of energy: Experimental observations. *Phys. Fluids*, **10**, 3126–3136.

Parker, E. N. 1957. Sweet's mechanism for merging magnetic fields in conducting fluids. *J. Geophys. Res.*, **62**, 509–520.

Parker, E. N. 1958. Dynamics of the interplanetary gas and magnetic fields. *Astrophys. J.*, **128**, 664–676.

Perez, J. C., and Boldyrev, S. 2008. On weak and strong magnetohydrodynamic turbulence. *Astrophys. J. Lett.*, **672**, L61–L64.

Pétrélis, F., Fauve, S., Dormy, E., and Valet, J.-P. 2009. Simple mechanism for reversals of Earth's magnetic field. *Phys. Rev. Lett.*, **102**(14), 144503.

Petschek, H. E. 1964. Magnetic field annihilation. *NASA Special Publication*, **50**, 425.

Plunian, F., Stepanov, R., and Frick, P. 2013. Shell models of magnetohydrodynamic turbulence. *Phys. Rep.*, **523**, 1–60.

Podesta, J. J., Roberts, D. A., and Goldstein, M. L. 2007. Spectral exponents of kinetic and magnetic energy spectra in solar wind turbulence. *Astrophys. J.*, **664**, 543–548.

Poincaré, H. 1889. *Sur le problème des trois corps et les équations de la dynamique.* Text presented to the Swedish Royal Academy.

Politano, H., and Pouquet, A. 1998. Von Kármán–Howarth equation for MHD and its consequences on third-order longitudinal structure and correlation functions. *Phys. Rev. E*, **57**, R21–R24.

Ponomarenko, Y. B. 1973. Theory of the hydromagnetic generator. *J. Appl. Mech. Tech. Phys.*, **14**, 775–778.

Ponty, Y., and Plunian, F. 2011. Transition from large-scale to small-scale dynamo. *Phys. Rev. Lett.*, **106**(15), 154502.

Pouquet, A., Frisch, U., and Leorat, J. 1976. Strong MHD helical turbulence and the nonlinear dynamo effect. *J. Fluid Mech.*, **77**, 321–354.

Priest, E. 2014. *Magnetohydrodynamics of the Sun.* Cambridge University Press.

Ravelet, F., Berhanu, M., Monchaux, R. *et al.* 2008. Chaotic dynamos generated by a turbulent flow of liquid sodium. *Phys. Rev. Lett.*, **101**(7), 074502.

Richardson, L. F. 1922. *Weather Predictions by Numerical Process.* Cambridge University Press.

Rikitake, T. 1958. Oscillations of a system of disk dynamos. *Proc. Cambridge Phil. Soc.*, **54**, 89.

Roberts, G. O. 1970. Spatially periodic dynamos. *R. Soc. Lond. Phil. Trans. Series A*, **266**, 535–558.

Rogers, B. N., Denton, R. E., Drake, J. F., and Shay, M. A. 2001. Role of dispersive waves in collisionless magnetic reconnection. *Phys. Rev. Lett.*, **87**, 195004.

Saddoughi, S. G., and Veeravalli, S. V. 1994. Local isotropy in turbulent boundary layers at high Reynolds number. *J. Fluid Mech.*, **268**, 333–372.

Saur, J., Politano, H., Pouquet, A., and Matthaeus, W. H. 2002. Evidence for weak MHD turbulence in the middle magnetosphere of Jupiter. *Astron. Astrophys.*, **386**, 699–708.

Schekochihin, A. A., Cowley, S. C., Dorland, W. *et al.* 2009. Astrophysical gyrokinetics: Kinetic and fluid turbulent cascades in magnetized weakly collisional plasmas. *Astrophys. J. Suppl.*, **182**, 310–377.

Schmidt, W., Federrath, C., and Klessen, R. 2008. Is the scaling of supersonic turbulence universal? *Phys. Rev. Let.*, **101**(19), 194505.

Servidio, S., Matthaeus, W. H., Shay, M. A., Cassak, P. A., and Dmitruk, P. 2009. Magnetic reconnection in two-dimensional magnetohydrodynamic turbulence. *Phys. Rev. Lett.*, **102**(11), 115003.

She, Z.-S., and Leveque, E. 1994. Universal scaling laws in fully developed turbulence. *Phys. Rev. Lett.*, **72**, 336–339.

Shebalin, J. V., Matthaeus, W. H., and Montgomery, D. 1983. Anisotropy in MHD turbulence due to a mean magnetic field. *J. Plasma Phys.*, **29**, 525–547.

Shrira, V., and Nazarenko, S. (eds.) 2013. *Advances in Wave Turbulence*. World Scientific.

Sorriso-Valvo, L., Marino, R., Carbone, V. *et al.* 2007. Observation of inertial energy cascade in interplanetary space plasma. *Phys. Rev. Lett.*, **99**(11), 115001.

Spitzer, L. 1962. *Physics of Fully Ionized Gases*. Interscience.

Steenbeck, M., Krause, F., and Rädler, K.-H. 1966. A calculation of the mean electromotive force in an electrically conducting fluid in turbulent motion. *Z. Natur.*, **21**, 369–376.

Stieglitz, R., and Müller, U. 2001. Experimental demonstration of a homogeneous two-scale dynamo. *Phys. Fluids*, **13**, 561–564.

Strauss, H. R. 1976. Nonlinear, three-dimensional magnetohydrodynamics of noncircular tokamaks. *Phys. Fluids*, **19**, 134–140.

Sweet, P. A. 1958. The neutral point theory of solar flares. Pages 123–134 of Lehnert, B. (ed), *Electromagnetic Phenomena in Cosmical Physics*. IAU Symposium, vol. 6.

Tarduno, J. A., Cottrell, R. D., Davis, W. J., Nimmo, F., and Bono, R. K. 2015. A Hadean to Paleoarchean geodynamo recorded by single zircon crystals. *Science*, **349**(6247), 521–524.

Thomson, J. J. 1897. Cathode rays. *The Electrician*, **39**, 104–109.

Turner, L. 1986. Hall effects on magnetic relaxation. *IEEE Trans. Plasma Sci.*, **14**, 849–857.

Velikhov, E. P. 1959. Stability of an ideally conducting liquid flowing between cylinders rotating in a magnetic field. *J. Exp. Theor. Phys.*, **9**(5), 995–998.

Vermare, L., Gürcan, Ö. D., Hennequin, P. *et al.* 2011. Wavenumber spectrum of microturbulence in tokamak plasmas. *Comptes Rendus Phys.*, **12**, 115–122.

Vlasov, A. A. 1938. On vibrational properties of an electron gas. *J. Exp. Theor. Phys.*, **8**, 291–318.

von Kármán, T., and Howarth, L. 1938. On the statistical theory of isotropic turbulence. *Proc. R. Soc. Lond. A*, **164**, 192–215.

Woltjer, L. 1958. A theorem on force-free magnetic fields. *Proc. Nat. Acad. Sci.*, **44**, 489–491.

Yamada, M., Ren, Y., Ji, H. *et al.* 2006. Experimental study of two-fluid effects on magnetic reconnection in a laboratory plasma with variable collisionality. *Phys. Plasmas*, **13**(5), 052119.

Yamada, M., Kulsrud, R., and Ji, H. 2010. Magnetic reconnection. *Rev. Mod. Phys.*, **82**, 603–664.

Yokoyama, N., and Takaoka, M. 2014. Identification of a separation wave number between weak and strong turbulence spectra for a vibrating plate. *Phys. Rev. E*, **89**, 012909.

Zakharov, V. E., L'vov, V. S., and Falkovich, G. 1992. *Kolmogorov Spectra of Turbulence 1. Wave Turbulence*. Springer Verlag.

Zakharov, V. E. 1965. Weak turbulence in media with a decay spectrum. *J. Appl. Mech. Tech. Phys.*, **6**, 22–24.

Zank, G. P., and Matthaeus, W. H. 1992. The equations of reduced magnetohydrodynamics. *J. Plasma Phys.*, **48**, 85.

Index

Printed in the United States
by Baker & Taylor Publisher Services